# LANDS~

Landscape is a central theme of cultural geography, and the past twenty-five years have been an unprecedentedly fertile period for work in this area, with a series of different influential readings of landscape being debated and explored. Here, for the first time, distinctive traditions of landscape writing are brought together and examined as a whole, in a forward-looking critical review of work by cultural geographers and others. The book also presents readers with an understanding of landscape couched in terms of the tensions inherent in the concept. These tensions, the book argues, are creative and productive, provoking new agendas for geographical research on landscape.

Thus, this book is about the varied and sometimes competing and opposed ways in which cultural geographers and others have understood and defined landscape over the past twenty-five years. Chapter by chapter, it reviews and discusses examples of the types of research and writing that spring from different understandings and definitions of landscape. And in doing so it also highlights and examines the philosophical positions and critical and political agendas that underpin different understandings of and approaches to landscape. The book examines in turn empirical and materialist approaches to landscape, understandings of landscape as a 'way of seeing', work on cultures of landscape and recent landscape phenomenologies. Throughout, the interdisciplinary nature of landscape studies is also stressed, with work from art history, archaeology and visual and cultural theory discussed alongside that by cultural geographers. The final chapter focuses upon current trends and future prospects for cultural geographies of landscape.

*Landscape* is an advanced introduction to its topic. Student readers will find a thorough, informative and up-to-date account of one of the cardinal points of human geography. For researchers and lecturers, *Landscape* presents a forward-looking synthesis of hitherto disparate fields of enquiry, one which offers a platform for future research and writing.

**John Wylie** is Senior Lecturer in Cultural Geography in the School of Geography, Archaeology and Earth Resources at the University of Exeter.

# Praise for *Landscape*

This book synthesises earlier ideas and presents current thinking in an accessible form . . . an excellent contribution to the theoretical study of landscape.

Brian Short, University of Sussex

Very well written, very accessible, and easy to read quickly. A pleasure, in fact.

Richard H. Schein, University of Kentucky

# Key Ideas in Geography

SERIES EDITORS: SARAH HOLLOWAY, LOUGHBOROUGH UNIVERSITY AND GILL VALENTINE, LEEDS UNIVERSITY

The *Key Ideas in Geography* series will provide strong, original and accessible texts on important spatial concepts for academics and students working in the fields of geography, sociology and anthropology, as well as the interdisciplinary fields of urban and rural studies, development and cultural studies. Each text will locate a key idea within its traditions of thought, provide grounds for understanding its various usages and meanings, and offer critical discussion of the contribution of relevant authors and thinkers.

**Published**

*Nature*
NOEL CASTREE

*City*
PHIL HUBBARD

*Home*
ALISON BLUNT AND
ROBYN DOWLING

*Landscape*
JOHN WYLIE

**Forthcoming**

*Scale*
ANDREW HEROD

*Environment*
GEORGINA ENDFIELD

# LANDSCAPE

John Wylie

Routledge
Taylor & Francis Group

LONDON AND NEW YORK

First published 2007
by Routledge
2 Park Square, Milton Park, Abingdon, Oxon OX14 4RN

Simultaneously published in the USA and Canada
by Routledge
270 Madison Avenue, New York, NY 10016

Routledge is an imprint of the Taylor & Francis Group, an informa business

Typeset in Joanna and Scala Sans by
Keystroke, 28 High Street, Tettenhall, Wolverhampton
Printed and bound in Great Britain by
Antony Rowe Ltd, Chippenham, Wiltshire

British Library Cataloguing in Publication Data
A catalogue record for this book is available from the British Library

Library of Congress Cataloging in Publication Data
Wylie, John (John W.)
Landscape / John Wylie.
p. cm.
Includes bibliographical references and index.
1. Landscape assessment. 2. Cultural landscapes. 3. Human geography.
I. Title.
GF90.W95 2007
304.2—dc22
2007001873

ISBN10 0–415–34143–4 (hbk)
ISBN10 0–415–34144–2 (pbk)
ISBN10 0–203–48016–3 (ebk)

ISBN13 978–0–415–34143–1 (hbk)
ISBN13 978–0–415–34144–8 (pbk)
ISBN13 978–0–203–48016–8 (ebk)

To Sarah, Lola and Rosa

# Contents

# Boxes

# Acknowledgements

This book emerges from nearly ten years spent studying and writing about landscape, and so lots of different people and institutions have played some part in its production, even if this has been an unwitting one. I first began to seriously think about issues around landscape in the course of doing my PhD, and I'm very much indebted to my supervisors Paul Cloke and Nigel Thrift, and more generally to the School of Geographical Sciences at the University of Bristol. I feel lucky to have been a graduate student in Bristol through the mid- to late 1990s, when lots of new ideas and approaches were in the air, and where theoretical and conceptual work was truly embraced and taken seriously. I was a member of a large graduate school, but especial thanks for support, inspiration and discussion through the years goes to Paul Harrison, J.-D. Dewsbury, Billy Harris, Mark Paterson, Katrin Horschelmann, Derek McCormack, Emma Roe and David Conradson.

Most of this book was conceived and written while I was working in the Department of Geography at the University of Sheffield, and I would like to thank Andy Hodson, Rob Bartram, Chasca Twyman, Megan Blake, Chris Clark, Peter Jackson, Jess Dubow, Tariq Jazeel, Deborah Sporton and Charles Pattie for all their support and collegiality over the years. Thanks must also go also to the cohorts of Sheffield students who took my final-year undergraduate module on Landscape and Exploration, and further thanks to graduate students Jenny Carton, Nissa Ramsay and Erena le Heron, who took a Master's course on Landscape and Nature in the year that I used it as a forum to explore and discuss the content and structure of this book. In October 2006 I moved to a new post in the School of Geography, Archaeology and Earth Resources at the University of Exeter, and I am deeply grateful to the School for generously providing relief from teaching and admin in my first months at Exeter; this considerably assisted the completion of this book.

When I began the task of writing this book, I felt I was well versed in the various literatures on landscape. But it has turned out to be very much a reading and studying process, and I think I am much more aware now of the nuances and complexities of different traditions of landscape scholarship than I was in late 2003, when I wrote the initial proposal. So I would like to thank Sarah Holloway and my erstwhile Sheffield colleague Gill Valentine, the series editors, for inviting me to submit a proposal on Landscape, and for offering me the rare opportunity to write a text on the geographical topic closest to my heart. Thanks also to Andrew Mould, Zoe Kruze and Jennifer Page at Routledge for their patience and encouragement through all the delays and redrafts. The attentive and at times strongly voiced criticisms of three manuscript reviewers have also, on this occasion, genuinely improved the final text.

Lastly I would like to thank the people I love most: my parents, my grandmother Mary Jones, my two daughters, Lola and little Rosa, and, most of all, my wife Sarah – even though she doesn't like the picture on the cover!

# 1

## INTRODUCTION

### 1.1 TENSIONS

Landscape is tension.

That much is clear from Paul Cezanne's painting, the one reproduced on the cover of this book. Take a look at it again now. In the final years of his life Cezanne painted over sixty images of this same scene; a view over fields to Mont Saint-Victoire, in Provence, in southern France. Some of these paintings are just abstract sketches, lines and smudges of colour. Others are more realistic, so to speak, with trees and buildings and fields given roughly the shape and definition they have in everyday perception. But many of the Mont Saint-Victoire images are like the one we see here: vivid, dappling and receding patterns of yellow and green, with the mountain itself there on the horizon, spectral, in the same grey-blue shades as the sky it is painted against, yet somehow clear and distinct.

The tension that animates Cezanne's landscape is one that has also recurrently haunted landscape studies in cultural geography. It is a tension between proximity and distance, body and mind, sensuous immersion and detached observation. Is landscape the world we are living in, or a scene we are looking at, from afar? Alternatively, we could put this question in the following way: does the word landscape describe the mutual embeddedness and interconnectivity of self, body, knowledge and land – landscape as the world we live in, a constantly emergent perceptual and material milieu? Or is landscape better conceived in artistic and painterly terms as a specific cultural and historical genre, a set of visual strategies and devices for distancing and observing? This book, it should be noted, does not aim to resolve this tension, or to arrive at some definitive resolution regarding landscape. Instead it sets out to document how the tensions which animate landscapes have proved enduringly creative and productive for cultural geographers and others interpreting and writing about landscape.

Thus this book is about the various, sometimes competing and opposed, ways in which cultural geographers in particular have understood and defined landscape over the past thirty-odd years. Chapter by chapter, it reviews and discusses examples of the type of research and writing that spring from different understandings and definitions of landscape. And in doing so it also highlights and examines the philosophical positions and critical and political agendas that underpin different understandings of and approaches to landscape. In this way, the book is grounded primarily in debates that have occurred within cultural geography, but at the same time it aims to illustrate how these debates have often had a distinctively interdisciplinary flavour. Cultural geographers have both drawn upon and contributed to different and competing notions of landscape emanating, for instance, from art history, visual theory, anthropology and literary studies. And so, with reference to all of this, tension seems doubly apt as an opening gambit: not only is landscape precisely and inherently a set of tensions; there are also significant tensions and differences between the various traditions of landscape enquiry. Later in this introductory chapter I will map this out in a little more detail, through outlining the overall structure of the book. To begin with, however, and starting with the example of Cezanne's painting, I want to work through some of the tensions that go into the making of landscape.

## Tension 1: Proximity/distance

Once, in a letter to a friend, Cezanne wrote that 'the landscape thinks itself in me . . . and I am its consciousness'. This could be read on one level as artistic egotism, claiming unique and visionary insight into landscape, and also in a sense ownership of it. However, taking up these words quite differently, the philosopher Maurice Merleau-Ponty (1969) argued that Cezanne's art was testimony to a sort of originary and inescapable involvement, the artist plunging into landscape, as it were, with his whole body, and reaching a point where self and landscape fold together and even fuse. For Merleau-Ponty the phenomenologist, Cezanne's perception of landscape was exemplary. It became the visual expression of his own argument that observer and observed, self and landscape, are essentially enlaced and intertwined, in a 'being-in-the-world' that precedes and preconditions rationality and objectivity. Cezanne is not a detached spectator – his gaze enters the landscape, is entered by landscape. In lived, embodied experience eye and land rest in each other's depths, and when we in turn gaze upon this painting we see both at once: the painter's vision *and* the visible landscape, imprinted on each other. In some ways, for Merleau-Ponty, Cezanne had captured the immersive and tactile realities of our everyday perception; but beyond this he testified more generally to the principle that landscape names a *perceiving-with-the-world*. Thus, he argued, Cezanne's art sought 'to make visible how the world touches us' (Merleau-Ponty, 1969, p.244).

But a very different story can be told about landscape. For the cultural historian and literary critic Raymond Williams (1985, p.126), *'the very idea of landscape implies separation and observation'*. The argument here is that, far from being about tactility or proximity, landscapes set us at a distance. They turn us precisely into detached spectators, and the world into distant scenery to be visually observed. As an artistic genre and as a culturally conditioned habit of visual perception, one arguably unique to European and Western societies, landscape is a particular way of seeing and representing the world from an elevated, detached and even 'objective' vantage point. It can be thought of as akin to other visual technologies (microscopes, telescopes, sextants) and modes of representation (cartography, architectural drawing) in which the world is conceptualised as an external, separate reality to be rationally perceived and accurately represented. Landscape thus belongs to science, rationality and modernity; it is the accomplice and expression of an epistemological model whose central supposition posits a pre-given external reality which a detached subject observes and represents.

Cezanne's painting of Mont Saint-Victoire might appear to have little in common with rationality and objectivity. In place of the formal and ordered conceptions of much classical landscape art, organised by the geometrical key of linear perspective, he offers what seems an unforced, intense and organic perception of a landscape itself already resonant and vibrant. But, this is still landscape in the 'scenic' sense; you still have to hold the book at a distance and measure your look, you still have to, as it were, stand back and get things into perspective. In other words, we can argue that Cezanne's art must still be understood in the broad cultural context of modern systems of visual ordering, systems which enshrine a certain *distance* (between viewer and painting in the art gallery, for instance), and which associate seeing with knowing, subjectivity with attentiveness, distance with authority, the mind with dispassionate contemplation. The visual theorist Jonathon Crary (2000) makes an argument along these sorts of lines with respect to Cezanne's art. He rejects Merleau-Ponty's implicit claim that Cezanne had somehow found a way back to an unvarnished perception of the world; in fact he rejects all attempts to view works of art as 'timeless', that is as having meaning outside of specific historical and cultural movements and agendas. Instead, for Crary, Cezanne's landscapes are to be understood in the context of a crisis of perception and specifically attention in the late nineteenth century. What we see in these landscapes is a recognition that 'looking at one thing intently did not lead to a fuller and more inclusive grasp of its presence, its rich immediacy. Rather it led to its perceptual disintegration and loss' (Crary, 2000, p.288). This, Crary argues, is a deeply historical moment in the development of new cultural and technical forms of vision, an 'interregnum suffused with possibilities, after the uprooting of vision from the classical order of knowledge in the seventeenth and eighteenth centuries but before its thorough relocation in regimes of machine vision which take off in the twentieth century' (ibid., p.359). Cezanne's landscape, in other words, sits squarely, albeit tensely, within an overarching historical narrative in which Western visual cultures are implicated in issues of order, power, spectacle and control. To visualise is to set at a distance.

## Tension 2: Observation/inhabitation

Is landscape a scene we are looking *at*, or a world we are living *in*? Is landscape all around us or just in front of us? Do we observe or inhabit landscape?

This distinction might seem a specious one – it can easily be argued that observing and inhabiting are not mutually exclusive, in so far as we are always doing both. Or we could simply say that looking is part of living, not an adjunct to it. Alternatively, we could turn once again to Cezanne's painting of Mont Saint-Victoire, and this time note that his power of observation is so intense that he seems to be *inside* the landscape he is painting, living and breathing with it, so to speak. Here, therefore, observation in a way *becomes* inhabitation. When we observe or visually explore something intently, we do indeed, contra Crary's arguments above, sometimes have the sense of getting closer to it, sometimes even of getting inside it.

Nonetheless, a distinct tension between observation and inhabitation often structures landscape. In particular, I would suggest, these two words have stood and continue to stand for different ways of studying and knowing landscape. For much of the twentieth century, for example, a broad 'field sciences' model underwrote the study of landscape by human geographers. As will be discussed in Chapter 2, geographers such as Carl Sauer and landscape historians like W.G. Hoskins based their work above all on first-hand observation, on observation-in-the-field. In a way, their ability to speak authoritatively about landscape was predicated upon their having lived it, touched it, explored it on foot – inhabited it, that is. But, at the same time, while they might have been passionate advocates for the value and worth of specific landscapes, and while they might have spent extended periods immersed in the field, much of Sauer's and Hoskins's writing – and much of its power – is rooted in a particular mode of neutral, empirical observation. They are *expert observers*: they stand back and take a detached view, and so they are able to bring together a wide-ranging synthesis. Close, careful observation leads to the accumulation of indisputable factual evidence. Trends and features are recorded and presented. Under the aegis of a certain science of observation, landscape takes shape as an external, syncretic, observable whole.

Much work on landscape in human geography and cognate disciplines since the 1970s has sought to move away from a 'field science' model, and has chosen to emphasise instead the *qualities* of landscape: landscape as a milieu of meaningful cultural practices and values, not simply a set of observable material cultural facts. Yet, here again, tensions between observation and inhabitation as modes of classification and enquiry continue to reverberate. On the one hand, a strand especially influential within cultural geography has moved towards the interpretative methodologies of the arts and humanities (see Chapters 3 and 4). Here, landscape is considered historically and culturally as a particular visual mode of observing and knowing. And the tone of academic enquiry is itself mostly observational – the accent is upon informed critical examination and dissection; the stance, while no longer neutral or remotely 'scientific', is nevertheless set at a distance from landscape, where landscape is understood as the vehicle of vested interests and regimes of power, and where the task of the critical cultural geographer is to interrogate such regimes.

On the other hand, a different set of approaches, broadly phenomenological in character (see Chapter 5) and influential in interpretative archaeology and cultural anthropology, as well as, increasingly, UK-based cultural geography, has sought to position the cultural practice of landscape very much within notions of embodiment, inhabitation and dwelling. Here, one argument is that a narrowly observational field science misses altogether the everyday textures of living and being in landscape – misses, in other words, the point of view of a landscape's *inhabitants*. To access this point of view, it has been argued, the researcher must not only theorise landscape via corporeal dwelling, but also come to know landscape through participating in it with his or her whole body (see Tilley, 2004).

A tension so threads through the couplet observing/inhabiting, because it signals different approaches to the study of landscape – different epistemologies. And thus a gap opens up in contemporary landscape studies, a gap between observing and inhabiting, between the critical interpretation of artistic and literary landscapes and the phenomenological engagement of cultural landscape practice. Chapter 5 in particular will examine the differences between these approaches to landscape.

## Tension 3: Eye/land

Contemporary English dictionaries commonly define landscape in something like the following terms: '*that portion of land or scenery which the eye can view at once*'. Most then go on to note that the term landscape may refer to a picture or image of the land, as well as the land itself. Finally, there is usually a reference to the verbal form, 'to landscape' – in other words to physically shape, modify and maybe even improve the land.

What do these, as it were, common or customary understandings tell us about the meaning of landscape? First of all, there is a clear suggestion that, as a 'portion of land', landscape belongs to an external, objectively real world. Landscapes are real; in other words, they are really out there: solid, physical and palpable entities, and not just figments of the imagination. Landscapes are the topographies we see and the terrains we travel through: the fields and the cities and the mountains. They may be surveyed, mapped and described in a factual and objective manner. This notion of landscapes as material and durable entities, as measurable and reliable records of both past and contemporary processes, is, of course, one that has historically been very important for geography, and also for linked disciplines such as archaeology and landscape history itself.

Second, however, the dictionaries state that a landscape is 'scenery' – something viewed by an eye; that is, by one, individual person. A landscape is thus not just the land itself, but the land as seen from a particular point of view or perspective. Landscape is both the phenomenon itself *and* our perception of it. In other words, while being linked in one way to what are usually called objective facts, to the real world 'out there', landscape is also found in the eye of the beholder. That is, landscape takes shape within the realms of human perception and imagination.

We can put this another way: landscape is not only something we see, it is also a *way of seeing things*, a particular way of looking at and picturing the world around us. Landscapes are not just about *what* we see but about *how we look*. To landscape is to gaze in a particular fashion. And how we look at things is not only to do with the biological functioning of our eyes. How we look at things is a cultural matter; we see the world from particular cultural perspectives, the ones into which we have been socialised and educated. What this means – and this has become almost axiomatic for cultural geographers – is that studying landscape involves thinking about how our gaze, our way of looking at the world, is always already laden with particular cultural values, attitudes, ideologies and expectations.

Thinking on these lines we can turn towards the second element of the dictionary definition of landscape: a *picture or image of the land*, in other words a work of art. Over the years I have spent teaching undergraduate geography in the UK, I have routinely found that, when I ask new students what they would immediately associate with the word 'landscape', the most common answer is a picture, or a painting. It may well be that this is itself an example of cultural specificity – the same question addressed to a group of US students would likely produce a different response, colleagues tell me; landscape might well be thought of primarily in terms of land-shaping activities such as gardening or architecture. Nonetheless a pictorial understanding of landscape has been extremely influential (and not least in gardening and architecture), and is still reproduced unremarked in many places, for example in the Microsoft Word™ program I am using to write these words, where the page can be oriented as either 'portrait' or 'land-scape'. Perhaps above all, then, as the cultural geographers Denis Cosgrove and Stephen Daniels (1988) argued in their influential introduction to *The Iconography of Landscape*, landscape is an artistic genre, a style of painting, one originating, as many authors have described (see for example Cosgrove, 1985), in fifteenth-century Italy, with the invention of new perspectival techniques for representing space and depth on the canvas. And through the centuries artistic landscape forms have evolved and developed, becoming associated in particular with the depiction of rural and natural scenes, with the portrayal of nationally significant landscapes, and with aesthetic discourses such as the pastoral, picturesque and sublime.

The point here is that, clearly, landscapes are human, cultural and creative domains as well as, or even rather than, natural or physical phenomena. Landscapes are cultural representations, they are works of landscape art, paintings, photographs, descriptions in novels and travel guides. This is the understanding of the word landscape at work within subjects such as art history and visual and aesthetic theory; and, again, as we will see throughout this book, this is also an understanding of landscape that has come very much to the fore in cultural geographies since the mid-1980s.

So it would seem that the choice is squarely *eye* over *land*, subjective perception over objective entity. But in practice most people want to have both, or want to claim that subjective and objective both exist, albeit in different ways. We perform this division all the time. We assume, on the one hand, the existence 'out there' of an objective, phenomenal and material landscape of facts and figures, slopes and rocks and motorways and other measurable processes. Then, on the other hand, we acknowledge a subjective, perceptual and imaginative landscape composed of ideas, dreams, signs and symbols, cultural values, conflicting viewpoints, artistic conventions and so on. We split landscapes in two, in other words, we divide them up into 'material' and 'mental' aspects, objective and subjective, science and art, nature and culture.

Within cultural geography, the tension between eye and land has been most clearly evident through persistent anxieties over the issue of the *materiality* of landscape (see Chapter 4, Section 4.2, and Chapter 6). While very few would wish to query or turn away from the critical purchase and kudos gained from identifying the geographical concept of landscape with the representation of landscape in art and literature, a sense sometimes surfaces that this 'dematerialises' landscape, taking us away from both 'the real world' and from traditions of geographical scholarship rooted in the study of landscape as a plaintively material and customary thing (e.g. Olwig, 2002).

Look again at the dictionary definition: 'that portion of land or scenery which the eye can view at once'. In here there is a tension that remains unresolved, and in so remaining in fact helps to make landscape something intriguing, creative and productive for academics, artists and writers. The definition does not attempt to arbitrate or distinguish between two landscapes, an external and internal one, a real and a perceived landscape. Instead it encompasses eye *and* land, observer *and* observed. Perhaps all landscapes, rural and urban, artistic and topographic, could be examined in terms of the tensions they set up and conduct between observer and observed, tensions between ways of seeing and interacting, and sensitive, material entities and processes. Sometimes these landscaping tensions might be distant and dispassionate ones, for example an aloof and detached or scientific visual gaze upon the land from afar, as is maybe the case with some types of writing, mapping, surveying and landowning, and also within some traditions of Western landscape art. Sometimes landscapes may be about the breakdown or absence of relations and communication; of course we can speak of landscape in terms of isolation, marginality or disenchantment. But sometimes the observer/observed couplet might be intimate and tactile and even spiritual or therapeutic.

## Tension 4: Culture/nature

Perhaps more than any other, the couplet culture/nature signals the tensions at work within the concept of landscape. Culture/nature relationships have often been taken to constitute the very heart of landscape studies in cultural geography, and in linked subjects such as anthropology and archaeology. Traditionally, as Chapter 2 will show, landscapes have been defined by geographers as the product of interactions between sets of natural conditions – weather, terrain, soil type, resources, etc. – and sets of cultural practices – agricultural practices, religious or spiritual beliefs, shared values and behavioural norms, the organisation of society vis-à-vis gender roles, property ownership and so on. Nature plus culture equals landscape in this account. What we witness when we examine landscape is a process of continual interaction in which nature and culture both shape and are shaped by each other.

Thus we arrive at the notion that distinctive national, regional or local landscapes are expressions of human responses to and modifications of natural environments over long periods – the sort of understanding of culture, difference and locality we might encounter in school geography or, to take a different example, in guidebooks for tourists. Countries like the UK and USA become patchworks or mosaics according to this understanding of culture–nature relationships: the Peak District and Lake District landscapes, the Deep South, the Appalachians, the Home Counties, the Midwest, the Highlands and Islands, the Norfolk Broads and so on, each of these being distinct physical environments that influence but also reflect the influence of distinct human cultures.

Since the 1970s, many cultural geographers have devoted much of their energies to critiquing and differently conceptualising this traditional account of culture–nature relations, and thence of landscape. For example, as might already be evident, the traditional distinction made between 'nature' and 'culture' as two wholly separable realms of existence in many ways merely rephrases the error of dividing landscapes up into two fields, into objective facts and layers of subjective meaning.

Thinking of nature and culture – of natural processes and human cultural practices and values – as distinct and independent realms is, cultural geographers have argued, extremely problematic in both theory and practice. The issue of where one draws the line between the two becomes fraught with political, moral and ethical dilemmas. Were humans once part of nature? If so, how and when did they separate themselves off from it? If so, does this mean that some human cultures are more 'natural' than others? Is nature then fixed and given and culture dynamic and pliable? Or are cultural practices simply responses to natural environmental conditions? At times in the past geographers *have* fallen into a 'natural' or environmental determinism, akin to and formative of racism, in which the physical environment – the climate, the soil, etc. – is seen as determining the cultural level of its inhabitants, in other words their state of social, political, artistic and technological advancement.

Partly in order to address these and other such issues, many cultural geographers have chosen to focus upon the complex cultural histories of the concept of 'nature'; that is the question of how certain images and understandings of nature have arisen via art, literature, science and philosophy, and thence become widely accepted and entrenched, not least in terms of actual environmental and development policies. Thus the cultural 'construction' of nature has become a key topic for geographers, and landscape imagery (as Chapters 3 and 4 will discuss) has been viewed as a key mechanism through which a particularly Western and European vision of culture–nature relations has been pictured and communicated.

But where does this leave nature, especially in terms of will, force and presence? Is nature a vital medium or inert, thoughtless matter? In focusing upon images of nature, landscape geographers, it has been argued, offer an overly culturalist or symbolic account, one in which nature becomes either little more than our ideas or perceptions of it, or is pictured as a blank canvas or screen on to which cultural meanings are projected. Perhaps it is unproductive to think of 'nature' and 'culture' as two primary, given terms whose interaction, materially or discursively, produces 'landscapes'. Perhaps instead we should think of *landscaping* first. That is, we should think about practices, habits, actions and events, ongoing processes of relating and un-relating, that come before any separation of 'nature' and 'culture'. Instead of landscape being the outcome of interactions of nature and culture, *practices* of landscaping – everyday things like walking, looking, gardening, driving, building – are in actuality the cause and origin of our ideas of what is 'nature' and what is 'culture'. Since the late 1990s, cultural geographers have begun to emphasise these sorts of approaches, most often termed phenomenological or performative, to issues of landscape, culture and nature. Chapter 5 of this book looks in particular at such work.

## 1.2  AIMS AND STRUCTURE OF *LANDSCAPE*

The previous section sought to advance an understanding of landscape in terms of a series of tensions – tensions between different understandings of the concept that are, nonetheless, creative and productive tensions, sparking renewed debates and continuing agendas for geographical research on landscape. Hopefully that section has given a flavour of what lies ahead. Here, I want to provide a more schematic outline of *Landscape*'s aims and structure.

This book focuses upon the different ways in which cultural geographers have conceptualised and studied landscape. It discusses some of the origins of landscape studies in human geography; for the most part, however, the text concentrates on landscape writing since the 1980s. In that time there have been many distinctive and often opposed and competing understandings of what landscape is, how it functions and what methods should be used to study it. In turn, these different understandings of landscape reflect the influence of different philosophical and political beliefs and agendas. They also reflect the ways in which cultural geographers both influence and are influenced by other disciplinary positions, and in the case of landscape the list of engaged academic subjects is long. So, in this book there are discussions of and references to work from history, art history, literary studies, anthropology, archaeology, critical theory, cultural studies and continental philosophy.

The aim here is to navigate a clear and plausible route through what are very diverse and fertile sets of writing about landscape. To enable this, the structure of the book is broadly chronological, moving gradually from the recent past to the early twenty-first century, as a means of highlighting and explaining change and progression. Within this framework, each chapter is a fairly self-contained examination of particular understandings of landscape and the key texts and concepts associated with them. This focus upon concepts and debates is offset by a series of 'case study' boxes, usually two or three per chapter (see List of boxes, p.ix), in which notable examples of the empirical and interpretative settings for particular approaches to landscape will be presented.

The book as a whole is best thought of as a forward-looking synthesis, which provides a contemporary consolidation and discussion of work on landscape, and hopefully provides a platform for future research and writing. With this agenda in mind, the final chapter (Chapter 6) seeks to discuss the prospects for landscape studies in cultural geography, focusing upon current and emergent trends, and this necessarily demands a slightly greater element of personal argumentation and reflection.

As its title implies, Chapter 2, 'Landscaping traditions' has a preliminary function. It sketches some of the major figures and tenets of landscape study through the twentieth century as a means of setting in context the more recent work with which the rest of the book is concerned. This historical reconnoitre also helps illustrate how and why 'landscape' has come to serve as both a substantive topic of study and a key organising idea within human geography. The chapter focuses in turn upon the North American school of cultural landscape analyses associated with Carl Sauer, the UK-based conceptions of landscape history epitomised by the work of W.G. Hoskins, and the 'vernacular' landscape traditions defined and promulgated by another American writer, J.B. Jackson. Beyond simply outlining the work and thought of such figures, the chapter seeks to show how they are united by a common sense that landscape may be defined in terms of an objective world of physical features, one that is 'out there' and that can be empirically accessed and described. This, as it were, commonsensical conception of landscape is further ramified by, first, an emphasis upon historical research, reconstructing past landscapes and tracing their evolution and, second, a general focus upon non-urban and more broadly pre-modern or pre-industrial spaces as privileged sites for landscape analysis. The chapter argues that it is in part important to be aware of such traditions because they constitute a set of precepts and arguments against which all subsequent generations of cultural landscape analysis either explicitly or implicitly define themselves. This is especially true in the case of Carl Sauer and the Berkeley School. It is also true, in a different way, of the extensive writings of J.B. Jackson. The chapter shows how these, while consonant with the work of Sauer and Hoskins in some respects, have nonetheless continued to provide inspiration for later schools, in particular the 'humanistic' geographers of the 1970s, and, more recently, a new generation of landscape phenomenologists.

Chapter 3 'Ways of seeing', focuses upon the revolution in landscape studies which occurred from the mid-1980s to the mid-1990s, under the aegis of the cultural turn in human geography. Here, as the title of the chapter implies, landscape was defined less as an external, physical object, or as a mixture of 'natural' and 'cultural' elements, and more as a particular, culturally specific way of seeing or representing the world. In this definition landscape is quite closely identified with landscape art, a complex and diverse artistic genre evolving from the fifteenth century to the present day. As a system for producing and transmitting meaning through visual symbols and representations, landscape art, alongside cognate arts such as cartography, photography, poetry and literature, is a key medium through which Western, and in particular European, cultures have historically understood themselves, and their relations with other cultures and the natural world. In adopting the concept of landscape as a way of seeing, new cultural geography thus emphasised the visual qualities of landscape, and also tended to focus upon *representations* of landscape in art, literature, photography and other media. Such representations, or cultural images, the argument runs, may be understood and analysed as expressions of cultural, political and economic power. The chapter is organised around three influential metaphors through which new cultural geographers sought to critically position and interpret landscape representations: landscape as veil, landscape as text and landscape as gaze.

Following on directly from Chapter 3, Chapter 4, 'Cultures of landscape', explores the diverse and diversifying geographical literatures which have emerged through the 1990s and up until the present day, in the wake of the initial formulations of new cultural geography. In part, the chapter argues, this is an evolutionary story: a story about how a central critical understanding of landscape as a way of seeing – landscape as a visual representation of cultural meaning and power – was extended, critiqued and developed by a second generation of cultural geographers. The notion of 'ways of seeing', with its critical imbrication of landscape, visuality, power and ideology, remains cogent today. But, at the same time, in work on 'cultures of landscape' past and present, and in analyses of landscape in the context of discourses of travel, colonialism and imperialism, a series of new and distinctive agendas and approaches open up. A shift from a broadly structuralist to a poststructuralist conception of vision, representation, knowledge, power and subjectivity is especially evident in recent landscape writing. The chapter approaches and discusses these shifts, first by considering debates around the question of the *materiality* of landscape. It then goes on to discuss, in turn, materialist landscape geographies mostly emanating from North America and writing on landscape, discourse and codes of self-conduct influential in a UK context and, finally, the now large and influential literature examining landscape, vision, representation and travel in the context of the discourses and practices of European colonialism and imperialism.

Chapter 5, 'Landscape phenomenology', describes and discusses an approach to landscape in some ways quite distinct from the broadly discursive and interpretative work considered in Chapters 3 and 4. This is the phenomenological approach, which, in the most general terms, eschews notions of landscape as an image, representation or gaze composed of specific cultural values and meanings, arguing that this enshrines and perpetuates a series of dualities – between subject and object, mind and body and, especially, between culture and nature. From a phenomenological standpoint, in contrast, landscape is defined primarily in terms of embodied practices of dwelling – practices of being-in-the-world in which self and landscape are entwined and emergent. The chapter's aim is to outline the major contours of this understanding, first by discussing the influential writings of Maurice Merleau-Ponty, the significance of whose thought for landscape studies in cultural geography and beyond lies most clearly in the manner in which he returns, time and again, to the entwined topics of vision and embodiment. A section of the chapter is then devoted to the work of the anthropologist Tim Ingold, which aims to explicitly rework landscape in terms of notions of dwelling and practice. Having established the terms of these key works on landscape phenomenology, the next substantive section of the chapter discusses the growing body of empirical work on various 'embodied acts of landscaping' (Lorimer, 2005) in cultural geography, noting this work's linkages with 'non-representational theory', and with emerging interdisciplinary studies of the body, perception, movement and materiality. The final section of the chapter, by contrast, discusses some of the main critiques of the phenomenological approach.

As already noted, Chapter 6, 'Prospects for landscape', is slightly different from the rest of the book, in that it looks to the future of landscape studies in cultural geography. With the potential pitfalls of forecasting in mind, the chapter limits itself to documenting what I understand to be the main conceptual and substantive issues currently preoccupying landscape studies in cultural geography. The first of these concerns the connections between landscape, identity and memory, and the chapter discusses how these have become a key substantive theme for landscape geographers in North America in particular since the late 1990s. Here, and in cognate UK-based writing, landscape is conceived in terms of struggle and conflict, and particularly so in debates over the material use and symbolic meaning of public landscapes, memorials and heritage sites. In part, the chapter argues, this reflects an understanding of landscape as the shared and disputed matter of daily life accrued from longstanding material and empirical traditions of enquiry. It also reflects the very influential role of forms of critical and radical politics, in which landscape is examined in terms of inequality and social justice. The next section of the chapter takes this idea forward, in outlining and discussing Kenneth Olwig's (2002) historical and lexicological re-casting of landscape as a political and legal entity. This re-casting, in which landscape is embedded within pre-Renaissance Northern European notions of community and custom, opens up new agendas for thinking through contemporary relationships between landscape and law. The following section, however, turns to consider recent, mostly UK-based, work being undertaken in the light of the non-representational theories discussed in Chapter 5. Work on the hybrid geographies of culture–nature relations is discussed in particular here, as this explicitly seeks to advance a relational, vitalist and topological vision of culture and nature, and space and society, in which notions of landscape would seem to have little purchase.

The final substantive section, by contrast, aims to illustrate the continuing cogency of UK-based landscape writing in the wake of non-representational theory, by discussing some examples of work that explores landscape via ideas of affect, presence, biography and movement.

## 1.3 CONCLUSION – LOOKING FORWARD

This chapter has sought to introduce the landscape concept by considering some of the major debates over its definition, meaning and use. It has thus defined landscape in terms of a series of tensions, and has argued that this is an important factor underlying the ability of landscape to persist as a cardinal term – a key idea – of cultural geography in particular. The tensions of landscape are creative and productive, in other words, rather than being indicative of some fault or lack in the concept, and hopefully this will become evident in the sheer diversity of the academic landscape research discussed in this book.

The chapter has also offered a schematic outline of the book as a whole. To reiterate, this book is about the various, sometimes competing and opposed, ways in which cultural geographers in particular have understood and defined landscape over the past thirty-odd years. Hopefully, in this way, it supplies a thorough and even-handed discussion, one that consolidates current work in a forward-looking synthesis, and thus provides a platform for future landscape writing.

# 2

## LANDSCAPING TRADITIONS

### 2.1 INTRODUCTION

The aim of this chapter is to take a retrospective look at some of the major figures and schools of thought within landscape studies in the twentieth century, up to about the mid-1970s. As such its work is preliminary and contextual. It sets out to discuss the intellectual contexts, sometimes from which, sometimes in reaction to which, subsequent understandings of landscape have developed. This may illustrate the continuing resonance of older landscaping traditions, because some, at least, have continued to provide both foundation and impetus for contemporary writers on landscape. In this way, the chapter is not organised solely around explicating primary sources. Instead, it looks quite consciously at the past from the perspective of the present, and uses more recent writings by cultural geographers as lenses for examining older schools of landscape analysis.

The chapter organises itself around three key figures in the history of Anglo-American landscape studies. The first of these is Carl Sauer (1889–1975), whose elaboration of an empirical cultural and historical geography, based upon the limpid observation and recording of material features in the field, formed the basis of the 'Berkeley School' of landscape studies. This school of thought, highly influential within academic geography in the USA from the 1920s to the present day, established landscape, and in particular *cultural landscape*, as a primary domain of analysis for human geography as a whole. In turn, Sauer's writings have experienced a sort of secondary after-life, a reprise in a minor key, because through the 1980s and 1990s they have been presented anew to students and researchers as a terrain against which 'new cultural geographies' have sometimes defined themselves (e.g. Duncan, 1980; Jackson, 1989; Mitchell, 1998b).

A second figure, but this time one with very much a UK basis, is W.G. Hoskins (1908–1992), whose best-known work, *The Making of the English Landscape* (1954), almost single-handedly defined and instigated a new sub-discipline of landscape history. In the process, landscape study and the concept of landscape itself became tightly linked to the study of rural local histories. As with Sauer, strands of continuing influence connect Hoskins with contemporary landscape work. But equally Hoskins's localist and empirical approach, while still influential in some branches of landscape history and landscape archaeology, has become quite distant from many contemporary cultural analyses of landscape.

The third and final figure in the landscape here is another American, J.B. Jackson (1909–1996). Never holding a formal academic position, Jackson nonetheless exercised a significant influence over generations of North American human geographers, cultural ecologists and planners, primarily as editor of, and leading contributor to, the journal *Landscape* from 1951 onwards. While he was committed, like Sauer and Hoskins, to an artefactual and empirical conception of landscape as an external reality in which material objects and features were distributed, two trajectories arrowing off in alternative directions are also present in his writing. First, there is a sense of landscape's avowedly cultural function as a source and repository of symbolic meaning and value (see Meinig, 1979a), a notion which, supplemented by a sharper critical analysis of relations of power and identity, comes to take centre-stage within new cultural geographies of landscape (see Chapter 3). Second, and perhaps more strongly, J.B. Jackson focuses quite distinctively upon the everyday or, as he puts it, 'vernacular' landscapes of post-war America, a world of garages, motels, mobile homes and highways, a world on the move and in the making rather than fixed or framed. In this way he may be revisioned as an early advocate of landscape understood in terms of experience, dwelling and embodied practice (see Cresswell, 2003) – an understanding which, via phenomenology and anthropology, has recently come once more to the fore in landscape studies (as Chapter 5 shall discuss).

## 2.2  CARL SAUER AND CULTURAL LANDSCAPE

### 2.2.1  Out in the open

A palette: on one side all burnt colours, red-browns, ochres, siennas, ter-racottas; on the other a glaring blue. The lines and colours of the open semi-arid landscape of the south-western United States, spare and rocky under the sky's bright bell. In a corner of this canvas a group of figures are sitting in a loose circle. One of them is talking, dominant, older; the others recline and listen.

The figure talking is Carl Ortwin Sauer, Professor of Geography at the University of California, Berkeley, for over thirty years from the 1920s to the 1950s, and effective originator of cultural landscape studies in the USA. I think it is useful to introduce Sauer in such a setting, a field setting, out in the open, a sort of open-air seminar room, because one of his signature beliefs was that fieldwork was the defining hallmark of both landscape studies and human geography more widely. For Sauer, fieldwork – getting out and about, preferably on foot with all senses alert, and as part of a group containing both experts and novices – defined both geographers and the landscapes they studied:

> Underlying what I am trying to say is the conviction that geography is first of all knowledge gained by observation. . . . In other words the principal training of the geographer should come, wherever possible, by doing field work.
>
> Such excursions and field courses are the best apprenticeship. The student and the leader are in running exchange of questions and promptings supplied from the changing scene, engaged in a peri-patetic form of dialogue about qualities of and in the landscape. . . . Being afoot, sleeping out, sitting about camp in the evening, seeing the land in all its seasons are proper ways to intensify the experience, of developing impression into larger appreciation and judgement.
>
> (Sauer, 1963b, p.400)

These words were spoken by Sauer towards the end of his career, as part of a valedictory presidential address to the Association of American Geographers in 1956. They nonetheless reveal much about his conception of landscape.

First of all they show that, for Sauer, landscape is very much to be found *out there*, beyond the lecture theatre and out past the city limits. Landscape, while *not* being synonymous with nature or wilderness (see below), is, nevertheless, here set at a clear distance from contemporary urban civilisation. Second, landscape is unequivocally real and solid, it is a physical material reality we are immersed in with all our senses. We cannot doubt the reality of the valleys, plains and farmlands we are camped out in. To put this another way, landscape is primarily factual and objective, it is an external, independent material field, a unified synthesis and arrangement of material forms and objects, and not a contrivance of our perception. We must therefore look outwards rather than inwards in order to understand it.

Third, the term landscape is here indistinguishable from, indeed is synonymous with, the term geography. Usually, landscape is simply *one* of the things studied by geographers, or is a term used in particular technical or specialist senses by geographers and others. But here, for Sauer, it takes on an expanded, holistic role. The landscape is geography. Geographers study landscapes.

A final point, and the most crucial point of all, is that when Sauer talks about landscape, he is talking about a *cultural entity*, something human-crafted, a modification of nature rather than a natural environment. Landscape is 'cultural landscape'. Geography is the science of cultural landscape. These understandings are set out most clearly in Sauer's (1963, originally published 1925) best-known programmatic statement, 'The morphology of landscape'. In this essay, he offers what has become a classic definition of landscape as the outcome of interactions between cultural and natural forces: 'the cultural landscape is fashioned from a natural landscape by a cultural group. Culture is the agent, the natural area is the medium, the cultural landscape the result' (1963b, p.343).

### 2.2.2  Landscape: *landschäft*, culture, morphology

In the 'morphology', Sauer claims that his understanding of culture, geography and landscape does not spring from intellectual reflection or specialist study. Rather, as an everyday phenomena, something we colloquially see and sense; landscape is already, self-evidently, the domain of geography, in the same way that plants and animals are the obvious subject-matter of biology. Sauer appeals to a domain of self-evident truths and common-sense understandings of reality as the basis for placing landscape at the centre of geography: this is something a child might well already know. 'Landscape is the field of geography because it is a naively given, important section of reality, not a sophisticated thesis' (ibid., p.316).

These precepts are expanded upon further later in the text, as follows:

> the term 'landscape' is proposed to designate the unit concept of geography, to characterise the peculiarly geographic association of facts. . . . Landscape is the English equivalent of the term German geographers are using largely, and strictly has the same meaning: a land shape, in which the process of shaping is by no means thought of as simply physical. It may be defined, therefore, as an area made up of a distinct association of forms, both physical and cultural.
>
> (Sauer, 1963b, p.321)

Aside from providing another expression of his essential 'nature plus culture equals cultural landscape' argument, what this quotation reveals is that, while Sauer was keen to present his understanding of landscape and geography as based on obvious and given facts, it was in actuality the product of an intellectual engagement with and synthesis of a number of different schools of thought. First, as signalled by Sauer himself in this quote, there are influences here from German geographical and anthropological traditions. In particular there is the notion that the concept of landschäft, in German simply referring to a 'bounded area' of land, could function as a basic unit of analysis and organising principle for the then still-emergent school and university subject of geography in the USA. It is worth noting here that landschäft itself is not a straightforward term; the Dutch equivalent, landschap, has a more strongly visual and artistic connotation, it is much closer to the idea of the land as perceived, or a picture of land, whereas the Germanic version ties much more closely to the land itself, to the notion of objective external space, and to words like 'area' and 'region' (see Jackson, 1984b). Further layering this semantic trail, the Old Dutch word landskab refers quite differently again to legal and administrative concepts of community, property and justice (see Olwig, 1996). However, Sauer very much advocated the 'bounded area' definition, in the process rather isolating geographical landscape studies from both artistic and legal-political arenas. It is worth noting that, coming himself from Germanic roots, Sauer had spent five years studying in Germany (Jackson, 1989).

But perhaps the key feature of the arguments presented in Sauer's 'morphology of landscape' is the emphasis placed upon the role of culture in shaping landscape, or, as the title of his much later essay put it, 'the agency of man [sic] on the earth' (Sauer, 1956). Leaving to one side for the moment issues around the definition of the term 'culture', Sauer's stress upon people's ability to not only adapt to but to shape and change their natural, physical environment arose in part from a dissatisfaction with the then paramount geographical discourse of environmental determinism. This school of thought, associated with North American geographers such as Ellen Semple and Ellsworth Huntingdon, and with links to both Darwinian evolutionary theory and colonial climatic science and physiography (see Livingstone, 1992), took its cue from the premise that 'man [sic] is a product of the earth's surface' (Semple, 1911, p.1). Hence it was concerned above all to trace and map environmental influences – influences of climate, terrain, soil, vegetation and so on – upon the development, evolution and migration of human cultures in various parts of the globe. In retrospect, given the ways in which it was complicit with the fashioning and per-petuation of racial stereotypes and hierarchies, the determinist paradigm does not appear as one of human geography's most laudable traditions. And in contrast Sauer's constant stress upon human agency and diversity is noteworthy for its attempt to reverse the direction of analysis, and to promulgate instead a cultural geography of how different landscapes came into being via both human and physical processes, and thence trace the pathways of active, rather than merely reactive, human cultures through landscape.

As both Jackson (1989) and Mitchell (1998b) argue, Sauer's insistence upon the effectivity and salience of human cultures vis-à-vis their inhab-itation and transformation of physical environments, may be traced intellectually back to nineteenth-century German Romanticism, and to figures such as Goethe and Herder, from whence ideas about the par-ticularity, value and vitality of certain 'cultural groups' (placed within certain 'cultural areas') such as indigenous and local cultures, first emerged. These constitute a set of nascent arguments which later culminate in twentieth-century European cultural nationalism. More directly in terms of empirical interests and methodology, Sauer's work on cultural landscape was influenced by cultural anthropologists such as Franz Boas and, in particular, a colleague at Berkeley called Alfred Kroeber, for whom the post-environmental-determinist challenge was to identify, describe and classify distinctive 'cultures' as they spread and distributed themselves over the earth's surface, distinguishing themselves through specific traits such as distinctive agricultural and architectural practices. To put this in Sauerian terms, the task was to describe the *morphology* – that is, the shape, form and structure – of a given landscape, and in so doing to reveal the characteristics, trace, distribution and effectivity of the human cultures that had inhabited and moulded it.

As Don Mitchell (1998b, p.27) aptly puts it, for Sauer, 'natural landscape was both the stage for, and the prime ingredient in, human geographic activity'. Distilling from these equations of culture, nature, landscape, is a certain ethical sense of harmony and fit, a sense that a 'good' landscape would exhibit humanity and nature in balance. Here, Sauer's ideas look at once backward, in an anti-urban, anti-modern, Arcadian and Romantic vein, and forward, in so far as they anticipate contemporary concerns with sustainability, and highlight the 'seeds of environmental disaster continually being sowed by believers in instrumental, technical rationality' (ibid., p.24). In practical terms, Sauer's disenchantment with the burgeoning industrial and urban culture of California led him further out into the less-developed wilds of the American south-west, further south into Latin America, and further into the past. In some respects, therefore, this was a cultural landscape geography more at home in the country and the past than the city and the present. More consequentially, in placing most emphasis upon morphology – upon facts and visible forms and records of past processes in non-urban, non-industrial landscapes – Sauer's work helped to align landscape studies, and much of academic geography in the United States, very much with the field sciences, with geology and botany especially, rather than with the then emergent social sciences such as sociology and psychology. A picture so emerges of landscape studies as the empirical description and discussion of the material traces – artefacts, patterns, settlements – left upon the land by bounded and coherent cultural groups.

## 2.2.3 The 'new cultural' critique of Sauer and the Berkeley School

Marie Price and Martin Lewis (1993) argue that it is misleading and even unfair to view Carl Sauer's work wholly through the prism of landscape and cultural geography. They highlight the range of his topical interests, and point out that, through the work of successive generations, he has given

**Box 2.1  CARL SAUER – 'THE PERSONALITY OF MEXICO'**

'This is an excursion into the oldest tradition of geography . . . the art of seeing how land and life have come to differ from one part of the earth to another.'

(Sauer, 1963a, p.104)

Originally published in the *Geographical Review* in 1941, 'The personality of Mexico' is an example of Sauer's holistic 'land and life' approach to geography, an essay blending together issues of climate and physical topography with cultural and economic history, aiming to produce a unified picture of the Mexican landscape. The essay gives a flavour of Sauer's commanding and synoptic writing style, also of his characteristic phrasing of 'cultural hearths' and 'traits'. And here the notion of 'personality' is used to express his influential vision of landscape as a dynamic relation of nature and culture:

The designation 'personality' applied to a particular part of the earth embraces the whole dynamic relation of land and life. It does not deal with land and life as separate things, but with a given land as lived in by a succession of peoples, who have appraised its resources . . . who have spread themselves through it as best suited their ends, and who have filled it with the works that best expressed their particular way of life.

(Sauer, 1963a, p.104)

Sauer's hypothesis is that Mexico is divided, physically and culturally, between an arid or semi-arid Northern upland (Chichimeca) and a more temperate, coastal and fertile Southern lowland. This Southern landscape, he further argues, is the 'cultural hearth' of Mexico – its birthplace and focal point.

The ruder cultures of the North occupied the interior tableland as far south as the base of the central volcanic chain. The advanced cultures held two great prongs extending northward in the coastal lowlands and foothills. . . . On the west coast, the extension of high culture (which I had the good fortune to discover a dozen years ago) reached into northern Sinaloa. In the west also, 'islands' of intermediate cultures . . . formed links to the Pueblo country of our Southwest. In general the expansive energy of the high cultures was notably greatest in the west, next greatest on the east coast, and least in the centre.

(p.106)

The origins of Mexican agriculture and hence culture per se are thence traced to the west (Pacific) coast:

the South and Southwest of Mexico constitute one of the great cultural hearths of the world, in which was created in part, and developed largely, an economic complex that is one of the great achievements of mankind. . . . Behind the named civilisations of the Maya, Aztec and Toltec lie older and more fundamental attainments in plant cultivation. . . . The basic traits of the native domesticated plants point to a source on the Pacific margin rather than the Atlantic. The Pacific areas have an in general shorter rainy season, a smaller total rainfall, and a much more sharply marked dry season. Their soils are rarely acid, most commonly they are somewhat alkaline. All the principal native crops show traits that point to an origin in the drier Western lands. Perhaps we may seek the earliest farming in western alluvial valleys.

(p.108)

And then, after some brief notes on metal use and political organisation in pre-conquest Mexico, Sauer moves on to the Spanish invasion and its consequences. He writes of the foundation of Nuevo Galicia in the west of Mexico, and its capital, Compostela, 'now drowsing around an ancient church that bears the double-eagles of Habsburg, it is remembered by us only because Coronado collected here the idle young gentlemen of New Spain to ride

continued

thence to the plains of Kansas' (p.114). At this point, in the mid-sixteenth century, industrial rather than agricultural history begins to take centre-stage.

> To the great good fortune of Spain it so happened that immediately behind the ramparts of the Maxton country lay the greatest silver country in the world – the land of the Zacatec indians. . . . Soon Zacatecas became the greatest silver producer in the world. In sustained production it has never been equalled by any other silver district. The Zacatecas strike was followed during the next quarter of a century by the discovery, without parallel in history, of a series of silver districts.
>
> (p.115)

> As wealth in unheard-of amount flowed south from the silver mines of the former Chichimeca, the native populations of the North were swept out of existence . . . Southern Indians were brought in as free labourers in an unending stream. . . . The richer mines imported droves of Negro slaves. Many Spaniards of small or no means came north to try their hand at mining, merchandising or transport of goods. In the course of time all these stocks except the upper-class Spaniards fused into a new breed, of no one colour. Thus was born the mestizo Mexico of today. Here was the frontier of New Spain, on which finally a new nationality was formed.
>
> This design of New Spain was drawn during the sixteenth century and has persisted to the present. Still the Northern march has dominance in part over the Southern hearth. It is still an area of immigration, receiving labour, foodstuffs and manufactured goods. . . . For the most part men of the North have made the revolutions and wielded the power. . . . The South still shows its aboriginal fundament of patient, steady toil done by apt craftsmen, who can still create things of remarkable beauty if they have the chance. The old line between the civilised South and the Chichimeca has been

blurred somewhat, but it still stands. In that antithesis, which at times means conflict and at other times a complementing of qualities, lies the strength and the weakness, the tension and harmony that make the personality of Mexico.

(pp.116–117)

impetus and direction to diverse fields such as cultural ecology, Latin American studies, 'people–environment' relations and types of vernacular historical geography. The argument is that any retrospective judgement upon an entire generation of academic activity will be a simplification and a caricature. Nevertheless, the fact is that the work of Sauer, his colleagues and students – the 'Berkeley School' – has, over the past twenty-five years or so, endured a strange sort of afterlife, particularly in the UK, in which it has been represented and to an extent repackaged as a set of hidebound and restrictive tenets, against which a new generation of cultural geographers have defined themselves and their work. A little oddly, many of these critiques have also emerged from the UK, where Sauer was never the figure of outstanding influence he was in North America. This has led to a situation in which most human geography undergraduates (myself included, as an undergrad a dozen years ago) now only ever encounter his work in a critical vein, and as a brief prelude to more in-depth examinations of contemporary cultural geographies and cultural politics.

Of course there *are* valid and pointed criticisms to be made of the Sauerian approach. One of the earliest, and still most often-cited, is voiced in James Duncan's (1980) critique of the 'superorganic' understanding of culture implicit in the work of Sauer and other early cultural geographers. As already noted, North American cultural geography has been historically angled towards the field sciences rather than the social sciences. Therefore, as two of its best-known promulgators, Wagner and Mikesell (1962), declared, it was especially and even primarily interested in visible, material 'evidence' of culture in the landscape – in the type and spatial distribution of built forms and artefacts – and not with the 'inner workings' of culture in terms of shared cultural beliefs, rituals, ideologies, values, attitudes and so on. In other words, and rather paradoxically, cultural geography lacked any real understanding or definition of culture, beyond the concept of 'superorganicism', in which culture is understood to be a uniform living entity independent of individual humans: 'by some remarkable process culture also lives on its own, quite apart from the single person' (Zelinsky, 1973, p.71).

What we see, therefore, in Sauerian cultural geography, is a *reification*, a process in which the central concept, culture, is isolated and put upon a pedestal. Culture comes to be viewed as a causative entity autonomous from but also determinative of individual attitudes and actions. The consequences for Sauerian cultural geography, as James Duncan (1980) names and elaborates them, are manifold and pernicious. First, the central concept, culture, is placed in abeyance as an unproblematic, given force, and is therefore not subjected to further theoretical scrutiny. Second, individuals are rendered passive and homogenous; they become merely the secondary bearers of cultural habits and styles and the tokens of cultural traits such as 'national character'. Finally, a totalising and determinist concept of culture precludes in advance any critical consideration of issues of difference and conflict. Thus the superorganic understanding of culture 'impedes explanation by masking many problematic social, economic and political relationships' (Duncan, 1980, p.198).

Pursuing a similar line of argument, Peter Jackson (1989, p.18) points out that by 'basing explanations in the transcendental realm of a suprain-dividual culture, [Sauerian] cultural geographers have often failed to address the wider social context in which cultures are constituted and expressed'. In other words, issues of social difference, social stratification, power, inequality and exclusion were inadequately conceptualised and addressed within this 'old' cultural geography. Therefore, what both Jackson and Duncan pointedly highlight is a comparative lack of a critical engagement with political and social issues within this form of landscape study – the very issues that many if not most contemporary writers would view as the essential ingredients and drivers of landscape (and as Chapters 3 and 4 shall discuss).

Two further criticisms flow from the inadequacies of a 'superorganic' understanding of culture. First, because of the lack of appreciation or theorisation of the implication of culture within social and political processes, the Sauerian cultural landscape was envisioned in a overly static and artefactual fashion. As Jackson (1989, p.19) summarises:

> Symptomatic of this approach is the [Sauerian] cultural geographer's almost obsessional interest in the physical or material elements of culture rather than its more obviously social dimensions. This focus

on culture-as-artefact has led to a voluminous literature on the geographical distribution of particular cultural traits from log buildings to graveyards, barn styles to gasoline stations. In contrast, much less consideration has been given to the non-material or symbolic qualities of culture or to other dimensions of the concept that cannot be 'read off' directly from the landscape.

Again the central issue here concerns the intellectual acuity and sophistication of the Berkeley School approach to cultural landscape. In privileging *objects* visible in the landscape, and, in some respects, being content to catalogue and map cultural traits in the landscape (log buildings, barns, etc.), those who carried Sauer's legacy forward promulgated a cultural geography that was able to *describe* but not *explain* landscape patterns and relationships, leading to an under-theorised and at worst shallowly empirical approach to landscape. The pedestrian fieldwork advocated by Sauer led to pedestrian scholarship, the argument runs, and as a consequence human geography lost a degree of prestige as an academic subject in the USA (on this, see Livingstone, 1992; Mitchell, 1998b).

Second, Jackson's indicative list of the sorts of things mapped and catalogued in the landscape – log buildings, barns, graveyards – serves to highlight the predilection of Berkeley School geographers for what Winchester *et al.* (2003) call 'the rustic and the exotic'. As earlier noted, Sauer's preference was for the rural and the wild, and for the great outdoors in general; his was a geography predicated upon a certain rejection of the urban, the modern and the mainstream. In consequence, although this is of course not without merit, the focus of much of his and his students' work was upon 'folk' cultures and traditions. And again, such a focus upon the small-scale and the rustic in the North American landscape served to position these cultural geographies of landscape as themselves self-consciously marginal to metropolitan intellectual currents.

To conclude this section, then, various critiques of Sauerian geographies are well established and in many ways well founded and, the intervention of defenders such as Price and Lewis (1993) notwithstanding, his conceptualisation of landscape itself is one that few cultural geographers today would wholeheartedly endorse. The 'superorganic' approach to cultural processes has come under sustained critique, as has the concomitant reliance upon a particular style of empirical field methodology of observing and describing. And yet, in a strange sense, these successive waves of critique have had the paradoxical effect of establishing the Sauerian vision of cultural landscape as the legacy with which all others must grapple, returning time and again to the empiricism, the fieldwork, the walking and looking and talking, the geology and vegetation and settlement patterns, the sheer materiality and presence of the landscape invoked. And Carl Sauer's landscape writing continues to resonate today, especially in North America (see for example Norton's (2000) introductory cultural geography text), and to appear, renewed, in unexpected places. Most recently of all, for example, his insistence upon the essential intertwining of geography and biology – of earth and life, as Whatmore (2006) puts it – appears prophetic once more in the light of presently emerging bio-geographies of culture and nature (e.g. Whatmore, 2002; Greenhough and Roe, 2006).

## 2.3   W.G. HOSKINS: LANDSCAPE, NOSTALGIA AND MELANCHOLY

### 2.3.1   Nostalgia

Is landscape always already nostalgic? Is there something inherently backward-looking about the concept, some quality which implies a turning-away from the present, a favorable contemplation of the past, a gaze upon a rose-tinted, evening-softened scene? Tim Cresswell (2003, p.269) voices a fairly common criticism when he writes that landscapes, and by implication landscape studies, are 'too much about the already accomplished', and that the word itself is 'altogether too quaint'. Already this chapter has noted Carl Sauer's identification of landscape studies with fieldwork in rural and remote locales and, thence, North American cultural geography's apparent preoccupation with rural backwaters. Turning to the UK, intellectual associations between the terms landscape, history, rurality and nostalgia are even stronger. The subject of this section, W.G. Hoskins – as Muir (1998) notes, probably the best-known British landscape historian – is someone who is perhaps remembered today as much for his hostile attitude to the post-Second World War world as for his academic scholarship. In the closing chapter of his celebrated study, *The Making of the English Landscape* (Hoskins, 1985 [1954] pp.298–299) Hoskins argues that:

> Especially since the year 1914, every single change in the English landscape has either uglified it, or destroyed its meaning, or both. . . .

Airfields have flayed it bare wherever there are level, well-drained stretches of land. Poor devastated Lincolnshire and Suffolk! And those long gentle lines of the dip-slope of the Cotswolds, those misty uplands of the sheep-grey oolite, how they have lent themselves to the villainous requirements of the new age! Over them drones, day after day, the obscene shape of the atom-bomber, laying a trail like a filthy slug upon Constable's and Gainsborough's sky. England of the Nissan hut, the 'pre-fab', and the electric fence, of the high barbed wire around some unmentionable devilment; England of the arterial by-pass, treeless and stinking of diesel oil. . . . Barbaric England of the scientists, the military men, and the politicians: let us turn away.

Today, Hoskins's anti-modern fulminations can seem a little reactionary. Yet the fact remains that he was undoubtedly an innovator in the field of landscape and local history studies in the UK. As such he can be plausibly placed beside Carl Sauer, not least as an incipient critic of the projects of Western modernity. Hoskins's words may appear backward-looking, but yet, as Matless (1993) notes, when set in context they are in actuality an early, inaugurating expression of what is a decidedly *contemporary* culture of nostalgia and melancholy with respect to landscape in the UK. Further to this, as Johnson (2006) observes, Hoskins's reaction to landscape has itself to be understood in terms of its wider cultural and intellectual inheritance of Romantic images of nature and landscape.

## 2.3.2 Temporality, locality, rurality

As in the case of Sauer, a series of schematic points may be made regarding Hoskins's conception of landscape. The first would be that, here, landscape is primarily the domain of *history* rather than geography. Hoskins was first and foremost an English historian. Born in 1908 in Devon, England, he spent the majority of his professional life working in history departments, at the universities of Oxford, Leeds and most notably at the Department of English Local History at the University of Leicester, which he founded (see Muir, 1998). *The Making of the English Landscape* thus approaches its topic not region by region but chronologically, charting evolutions, revolutions and continuities along the grooved track decreed by the conventional demarcations of British history (Celts, Romans, Dark Ages, Tudor and Georgian England, the Industrial Revolution), while spatially darting here and there. What emerges, therefore, is an intense sense of the *temporality* of the English landscape, the depth, richness and complexity granted by sheer cultural age. In this way Hoskins's conception of landscape is archaeological or geological rather than scenic. The landscape consists of vertical layers of use and inhabitation, it has been built upon over time. To understand it we have to excavate as much as gaze, because 'everything in the landscape is older than we think' (Hoskins, 1985, p.12).

This sense of the historical richness and detail potentially recuperable from any English scene is also in many ways the buttress of Hoskins's major intellectual contention in *The Making of the English Landscape*. His argument is against 'the wholly inadequate view that the English landscape is largely "the man-made creation of the seventeenth and eighteenth centuries"' (ibid., p.174), a view which he deems to have arisen as a consequence of the fact that those centuries witnessed both the rapid spread of open field enclosure and the flowering of classical landscape gardening in England. Time and again, Hoskins seeks to empirically counter this widespread assumption through historically evidenced examples of landscapes, as he puts it, 'completed' in Tudor or even medieval times.

This focus upon the cogency of the deep past in specific places means that Hoskins's historical account of landscape is also profoundly *localist*. His eyes are on the ground, on the details immediately to hand; they are not lifted to encompass the horizon and search for general rules and principles. The accent of *The Making of the English Landscape* is very much upon individual features: houses, bridges, fields, churches. In this way, in 'almost allying landscape and place *per se*', as Matless (1993, p.189) notes, Hoskins's project is one in which the term *landscape* blurs significantly with the terms *place* and *location*. The point which Matless goes on to draw from this observation is that Hoskins's emotional attachment is to an England of nooks and crannies, of hidden gems and forgotten coombes, a 'branch-line England':

> Hoskins' project could I think in many ways be termed a 'branch-line history'. This is a metaphor combining history and locality: it is also not without historical allusion. Hoskins' writings form part of a broader anti-modern, anti-state culture of landscape in England, one of whose expressions came in the surge of steam train and branch-line preservation from 1960 onwards. Like Hoskins' local study, branch-line salvage can find a virtue . . . in not connecting with a wider network, in operating locally, back and forth, and not beyond, in

an atmosphere of rescue and reverence for something felt to have
been passed by, bypassed.

(Matless, 1993, p.195)

Meinig (1979a) elucidates a related but more pointedly historiographic
point from Hoskins's localism, namely that his intellectual approach and
attitude ran very much against the then prevailing tendency of professional
historians to search for and focus upon general historical structures and
systematic forces in their analyses of societal change. Hoskins's localism
so represents a voice from an intellectual as well as geographical branch-
line, it is 'a stand at one extreme end of a spectrum' in relation to the practice
and purpose of historical research (Meinig, 1979a, p.203).The implication
is that 'landscape analysis is always a study of localities.The concern for factual
detail, the search for evidence visible in fields and hedges, lanes and streets,
buildings and clusters of buildings, sets all of his work within a certain scale.
For him landscape analysis . . . is necessarily a local form of history' (ibid.,
original emphasis).

As local history (and specifically English local history – the title of the
university department Hoskins founded as Leicester), Hoskins's vision of
landscape most clearly chimes with that of Carl Sauer in so far as it is rooted
in empiricism.The landscape is an objective, external, material assembly of
facts and things which is realised through direct encounter and observation.
In contrast to the archival and discursive tenor of much historiography, here
the principles and methods of the field sciences apply: knowledge is gained
primarily through getting out and about, again preferably on foot, and
being prepared to root around in neglected corners for the forgotten key
which might unlock an entire scene. Close, careful observation leads to
the accumulation of indisputable factual evidence.The landscape is a milieu
of solid fact rather than abstract theory. Its innate solidity and matter-of-
factness (it is made of bricks, fields, hedges, bridges), make it both an
indisputable object of enquiry and that which acts as a bulwark against any
sense of doubt and transience that a speculative or cosmopolitan perspective
might introduce.

As these words suggest, a fourth key element of Hoskins's conception
of landscape was rurality. Here, landscape is quintessentially countryside.
Of course The Making of the English Landscape includes chapters on 'the
Industrial Revolution' and 'the Landscape of Towns', however these are
largely presented as staging posts on a road to hell – 'the Landscape Today'.

Despite acknowledging 'a point . . . when industrial ugliness becomes sublime' (Hoskins, 1985, p.232), and presenting chemical-industrial Middlesbrough as a planned town alongside Romano-medieval Salisbury, the text tends to pinpoint rural arrangements – fields, farms, hamlets and even 'lively little English market towns' (ibid., p.297) – as a cultural and aesthetic ideal. The emphasis on detail, depth and locality becomes a means of enriching this ideal:

> For what a many-sided pleasure there is in looking at a wide view anywhere in England, not simply as a sun-drenched whole, fading into unknown blue distances, like the view of the West Midland plain from the top of the Malvern Hills, or at a pleasant rural miniature like the crumpled Woburn ridge in homely Bedfordshire, but in recognising every one of its details name by name, in knowing how and when each came to be there.
>
> (Hoskins, 1985, p.19)

'Anywhere' here might easily mean 'anywhere rural'. Hoskins's work may be read as a contribution to a long-standing and deep-rooted English discourse in which a certain rural idyll is represented as a source of aesthetic, social and ecological harmony. The key point here is that this idyll is understood to be always already vanishing. Indeed it can be argued that the threat of imminent destruction or obliteration is an integral and necessary feature of any such idyll. In this way 'tradition, beauty, meaning and history itself become confined in a past which is itself a refuge. . . . Hoskins sees the past as a source of solace and consolation, without the hope that it can be resurrected in the present. This country *can now only be in the mind's eye*, and so that mind must be melancholy' (Matless, 1993, pp.192–193, original emphasis).

### 2.3.3 Cultures of loss and melancholy

An irony here is that this pervasive melancholy and nostalgia, with its attendant idealisation of certain forms of landscape, may in some ways be clearly traced to artistic and literary discourses of picturesque and romantic landscape originating in the eighteenth and early nineteenth centuries (see Johnson, 2006). Hoskins's aesthetic and emotional claims regarding English landscape thus largely stem from the very period whose influence in making the *actual* landscape he wants to query. Nonetheless, a large measure of Hoskins's influence and success can be traced to his particular empirical and historical articulation of the rural idyll form. *The Making of the English Landscape* was, importantly, very much a popular as well as academic success. As Meinig (1979a) notes, its reception by professional historians was lukewarm, partly because of its 'field' and topographical basis, but partly also because of the emotional tenor of the writing. Yet it was this tenor, the open, declarative language of the text, that ensured a wide, non-specialist readership, and led eventually to both a commissioned television series, *One Man's England*, and an invitation from the poet John Betjeman, another important mourner for a vanishing, bucolic England, to author the influential Shell Travel Guides to Leicestershire and Rutland (see Box 2.2). Hoskins decried 'the unpalatable jargon of the geologist or the economic historian' (1985, p.19). Elsewhere he said 'I once wrote a book with the simple title of *The Making of the English Landscape*, but I ought to have called it *The Morphogenesis of the Cultural Environment* to make the fullest impact' (quoted in Meinig, 1979a, p.209).

---

**Box 2.2 W.G. HOSKINS – *THE SHELL GUIDE TO LEICESTERSHIRE***

'In the crowded Midlands, swarming with people and cars, heavily built-up and pylon studded, one is grateful for areas of space and quiet' (Hoskins, 1970, p.14).

Now collector's items evocative of a certain pre- and post-war English culture of landscape, the Shell Guides, produced erratically between 1934 and 1984, and edited for much of that time by the famous English poet John Betjeman, were a series of travellers' guidebooks intended for a newly mobile, domestic and educated English audience. Each guide took a particular English county (Cornwall, Shropshire, Norfolk, etc.), as its theme, and consisted largely of a gazetteer alphabetically listing items of interest to be found in that particular county. While aiming to provide a fairly comprehensive survey, these gazetteers often focused especially upon historical and architectural features appealing to

continued

the interested traveller, with churches, castles and country manors prominently described alongside major villages, towns and topographic landmarks.

W.G. Hoskins compiled the Shell Guides to the adjacent English Midland counties of Leicestershire and Rutland – this box focuses upon the Leicestershire text (Hoskins, 1970). It says something about the wide popular reach and scope of Hoskins's writing that he was commissioned by Betjeman to produce these guides. It is perhaps also worth noting that, at the time Hoskins was preparing this volume, the urban landscape of Leicester was being radically transformed by immigration from the Indian subcontinent and, in particular, the East African Asian diaspora. These processes, however, pass unremarked in the Shell Guide, which focuses squarely on the historical provenance and value of specific buildings, places and landscapes. Consisting of a gazetteer prefaced by an introductory essay, the Leicestershire guide is thus a good example of Hoskins's elegiac invocation of landscape and history; it also supplies several instances of his eye for idiosyncratic detail and his unapologetically trenchant judgements.

In this vein, Hoskins opens his introduction as follows: 'Leicestershire is generally dismissed by those who have merely driven through it on the A6 as flat, pretty well covered with red-brick towns and villages, with somewhere in the unseen background a lot of fox-hunting going on' (p.9).

Just as in his *magnum opus, The Making of the English Landscape*, the drawing-out of this 'unseen background', an historical/archaeological layer perceptible via the eye of the expert landscape historian, is the motivating goal of the guide to Leicestershire. We begin along the banks of the river Soar, dividing the county into an eastern and a western portion, though here 'much of it is unattractive at any time of year, a kind of half-derelict scruffy edge to the conurbation of Leicester, with some of the dreariest villages one could hope to see anywhere in Britain. There is something unbeatable about industrial red-brick put up in the decades between about 1880 and 1910. Even John Betjeman's heart would shrink at it, though there are moments of memorable ugliness suddenly encountered that redeem the general boredom' (p.9).

Hoskins's habitual anti-urbanism does not, however, altogether dismiss the charms of provincial Leicester; he happily pictures the city, for example, 'on sunny October mornings, when the Midland wind blows briskly along the residential roads, speaking of winter and the lecture season for the serious-minded' (*ibid.*). But soon the guide moves on to rural scenes, to village churches imbued with 'a pleasing air of Arnoldian decay. Melancholy on a grey winter day, shining like old gold on a summer evening, they are always a pleasure to look at' (p.12).

> I think especially of South Croxton and Tilton at any time of the year; but east Leicestershire is full of these treasures, often with a 17th-century manor hard by to complete the utter Englishness of the country scene. Much of High Leicestershire is over 600 feet up. The highest point is Whatborough Hill (755 feet above the grey North Sea), where the wind, so they proudly say in the East Midlands, blows straight from the Siberian wastes, with a 'lost village' on its very summit. No wonder the early village of Whatborough disappeared from that windswept plateau long ago – back in Henry VII's day, and decaying long before that.
>
> (p.12)

In the midst of this melancholy elegy, and moving west to Charnwood Forest, a surprising moment: 'in recent years the M1 has driven through it in a magnificent sweep, after much local opposition to the plan. I do not think the landscape historian can regret what he sees now. It is a fine road, and already a bit of history' (p.14).

But more that anything else, and in partial contrast to the patient historical reconstruction work that characterises his scholarly output, the Leicestershire guide gives Hoskins the chance to paint an earthier and riper picture of the landscape. The county, for instance, is famous for fox-hunting, 'it is the home of some of

continued

the most famous packs in England, perhaps the most famous – the Quorn, the Cottesmore, the Belvoir, the Atherstone, Fernie and Pytchley . . . I have been told that before the Second World War, a thousand fine hunters were brought to Melton at the beginning of the season, and the night air was sulphurous with aristocratic adultery' (p.16).

> Whatever one thinks of fox-hunting as a sport, it is one of the last survivals of the picturesque in England; it has produced some splendid literature; and it has dotted the English landscape with names that to their devotees cause a lifting of the heart – Billesdon Coplow, the Whissendine Brook, Ashby Pastures, Kirby Gate – but what can the layman know of these holy places?
>
> (p.16)

> There is one more remarkable thing about this small county: it has created three magnificent foods in its time . . . its Stilton Cheese, its Leicester Cheese, and its pork pies. . . . Since the early years of the 19th century there have been the wonderful pork pies one dreams about in exile. If you have not known the pork pies of the East Midlands, but only the factory-made product, you have never really lived, gastronomically speaking. . . . The red Leicestershire cheese was made in the southern part of the county in the 18th and 19th centuries. There were apparently several local cheeses under this general label, and Mr Monk, the reporter to the Board of Agriculture in the 1790's, thought some of these better than Stilton. Down to 1939, the best 'red Leicester' still came from the country south-west of Leicester, but the war and its rationing killed the farmhouse product, and the only genuine Leicester cheese is now made in a factory in Melton Mowbray. . . . As for Stilton cheese, named after a village on the old Great North Road (which is in Huntingdonshire) it is wholly a Leicestershire creation, despite the name attached to it. . . . In recent years the name Stilton

> began to be copied widely, like Cheddar or Champagne. The makers of the real thing fought a High Court action to protect their priceless name, and won it.
>
> Like so many things, Stilton is supposed to be not what it was before the war.
>
> (p.19)

Once again here a slightly intemperate note surfaces, this time directed at the specialist nature of academic vocabularies. Yet, as Matless (1993) stresses, when placed in the cultural and historical context of Britain in the early 1950s, Hoskins's work was novel and distinctive in so far as it ran against then dominant discourses of landscape planning and preservation, as epitomised by figures such as the geographer Dudley Stamp and the green-belt planner Patrick Abercrombie. Hoskins's anti-modern, anti-state, anti-urban voice is thus,

> an expression of a view of history, and of the relation between past, present and future which has informed much of the debate on conservation and heritage in England over the past forty years. It is important to stress, though, that in the late nineteen-forties and early nineteen-fifties there was a certain novelty in this view. Writers like Hoskins have been very successful in aligning ideas of conservation and landscape against modernity and 'progress'. But it was not always so. Inter- and post-war writers such as Stamp and Abercrombie and Thomas Sharp, all active in the cause of landscape preservation, did not turn from the modern world in disgust. . . . [T]heir preservation was a very 'progressive' and modern one . . . but in contrast to these progressive preservationists, Hoskins holds out no prospect of a contemporary beauty.
>
> (Matless, 1993, p.191)

W.G. Hoskins may thus be cast as an innovator in terms of helping to inaugurate a now commonplace vision of English landscape as a national nature and culture under threat and in large parts already destroyed. This melancholic narrative of loss and decline is nearly mainstream today, whereas the utopian modernism of the landscape 'planner-preservationists', with its credo of rational management, improvement and ordered techno-logical progress, has moved from being a dominant culture of landscape through the mid-twentieth century (see Matless, 1998) to a quite marginal discursive position.

In this way, 'Hoskins' turning away, his musing on loss, helped to make an emerging culture of English landscape from the nineteen-forties. England has been influenced by Hoskins, and his turning away from "The Landscape of Today" has, in its own way, made part of today's landscape' (Matless, 1993, p.201, original emphasis). Equally, *The Making of the English Landscape* was a pioneer work of landscape history, of landscape as history, and 'no-one has more consistently projected the reciprocal satisfactions of landscape analysis as a form of history and historical understanding as a form of landscape appreciation' (Meinig, 1979a, p.202).

Hoskins's influence thus continues to be felt, over fifty years after the publication of *The Making of the English Landscape*. As Johnson's (2006) recent text on ideas of landscape within the discipline of landscape archaeology highlights, Hoskins may himself be enrolled as a significant figure within a narrative of English landscape stretching back for centuries, to the Romantic movement and before. Equally, a conference at the University of Leicester in 2005, marking the fiftieth anniversary of the publication of *The Making of the English Landscape*, attracted a wide range of speakers from diverse academic disciplines. The majority, perhaps, had landscape archaeology and local history backgrounds, and, as continuing publications show (e.g. Thirsk, 2002; Muir, 2000), these are the academic areas in which Hoskins's inheritance is most clearly evident today. While his place within contempo-rary cultural geographies of landscape is more marginal, the conservationist and preservationist ethos he helped inaugurate has become a hallmark of rural policy and heritage policy in the UK – although what he might have to say about the actual contemporary condition of the English countryside is another question.

## 2.4 J.B. JACKSON AND 'VERNACULAR' LANDSCAPE

### 2.4.1 Inside and outside

Both W.G. Hoskins and Carl Sauer might be described as *insiders* with respect to the landscapes they studied, in so far as the work they produced, and in a way their entire approach to the study of landscape, was predicated upon an empirically close, even tactile, engagement with, and inhabitation of, the landscape. Their devotion to detail, their descriptive emphases, and in particular their undoubted level of emotional and intellectual investment in landscape gives rise in their writings to a sense of proximity, empathy and attachment.

But, on the other hand, Sauer and Hoskins might also be labelled *outsiders*. This is primarily because, as professional academics, both the conventions of scholarly writing and to an extent their own practice led them to present views of landscape from a more detached position; a synoptic overview, a factual record, a survey of the landscape as a whole by a knowing, if emotive, gaze. Their perspective is, in other words, that of the academic, even scientific, expert, someone who stands apart from the phenomenon in question, the better to objectively scrutinise it. As was discussed in the introductory chapter to this book, debates on landscape often pivot about this insider/outsider distinction. Is the landscape a picture we are looking *at*, from the outside? Or does the word refer to a world we are living *in*, a home or dwelling place?

The answer given by John Brinckerhoff Jackson, the final totemic figure to be discussed by this chapter, seems to immediately take one side. Jackson's stated credo is that 'far from being spectators of the world we are participants in it' (Jackson, 1997a [1960], p.2). His work thus seeks to articulate and advocate the insider's or inhabitant's point of view and, as I will detail below, to value in its own right the everyday contemporary or, in his words, 'vernacular' landscape. But at the same time one of the most interesting aspects of Jackson's writing is his blending of strongly voiced opinion with an informed cultural and historical awareness of the subtleties and ambiguities of the term landscape. This in itself has worked to ensure his work's continued cogency. As Cosgrove (1998, p.34) writes, Jackson 'opened out the concept of landscape in ways which seemed to democratise it, liberating both spectator and participant, by writing from the inside and pointing to the symbolic meanings which arise from social life in particular geographical settings'. For example, in his essay 'The word itself' (Jackson, 1984b), one of his more formal examinations of landscape, Jackson excavates in depth the etymological roots of the word, and demonstrates in this way the historic tensions between an insider's or inhabitant's understanding and the topographical, pictorial and scenic associations that have come to dominate usage as a consequence of centuries of landscape art, architecture and gardening. Elsewhere in the same volume of collected essays there are pieces on mobile homes, the symbolism of stone and 'landscape as seen by the military'. The body of writings bequeathed by Jackson is clearly eclectic and complex, it comprises 'a multidisciplinary and maverick conception of the study of landscape' (Cresswell, 2003, p.275), and in this section, in addition to outlining similarities and differences between Jackson, Sauer and Hoskins, I want to detail the ways in which cultural geographers in particular have adopted his legacy in sometimes quite different ways.

## 2.4.2  Vernacular landscape

Unlike Sauer and Hoskins, J.B. Jackson did not pursue a conventional academic career, that is, he did not work as a salaried university researcher and teacher, and this fact perhaps says much about his approach to and valuation of landscape. Born in France to American parents in 1909, he studied history and literature at Harvard, and subsequently served as an intelligence officer in the Second World War (Meinig, 1979a). In many ways, however, the key moment in Jackson's story arrived when, following the war, he leased a ranch in New Mexico in the south-west United States – the same region that had attracted and inspired Sauer. It was from this New Mexico base that he set out to pursue a distinctive career as an independent commentator on issues to do with American landscape; in particular, from 1951 onwards, as the publisher, editor and leading contributor to the journal *Landscape*. As Meinig (1979a, p.211) notes, this was in some ways an idiosyncratic and even quixotic venture, a voice from the margins in the form of 'an unheralded periodical issued from a post office box address in a small Southwestern city, with no hint of any kind of institutional or professional affiliation'. But from these obscure origins, regular issues of *Landscape* were subsequently published for over forty years, the journal expanding and diversifying through the years, and providing a unique forum for authors writing about landscape history, planning and ecology, issues of religion, myth and symbol, and, above all, the particularity and value of everyday, 'ordinary' places. The collected volumes of *Landscape* are thus in themselves testimony to J.B. Jackson's standing and influence, and they repay repeated browsing as a rich record of insight and commentary with respect to the design, use and meaning of landscapes.

The marginality of *Landscape*'s beginnings offers a point of entry to Jackson's own approach and stance. The constant and dominant theme of his writing is that the term 'landscape' refers to the material world of ordinary everyday life – the world of houses, cars, roads, sidewalks, backyards, the local, inhabited world of those who Jackson saw as ordinary Americans; people improvising and elaborating a life far removed from both metropolitan centres of power and abstract intellectual theorisation. A sense that this world – what he named as 'the vernacular landscape' – was being neglected, overlooked and even treated with disdain by academics, planners and politicians alike gave primary impetus and direction to Jackson's writing. Time and again in his essays for *Landscape* he performed the double manoeuvre of decrying specialist languages and remote perspectives while highlighting the richness and value of indigenous and vernacular forms of American life. In this way, above all else, as Tim Cresswell (2003, p.274) states, 'Jackson's landscapes are ones that people inhabit and work in and they are landscapes that people produce through routine practice in an everyday sense'. Jackson's own words are even stronger and clearer: 'we are not spectators; the human landscape is not a work of art. It is a temporary product of much sweat and hardship and earnest thought' (Jackson, 1997, p.343).

There are some echoes of Sauer and Hoskins here. All three writers want to eschew a purely aesthetic notion of landscape as beautiful scenery, and set themselves (and by implication landscape geography as a whole) at some distance from the formal and technical analyses of studies of landscape art. Here, art history is obviously if implicitly understood as an intellectual and even effete practice, one belonging to distinctively metropolitan, rarified spaces such as the gallery, museum and catalogue. The landscape described by Sauer, Hoskins and Jackson is defined by contrast in practical rather than aesthetic terms; it is, to use a word favoured by Jackson, 'workaday'. Equally, when Jackson (1995, p.43), in one of his final essays, defines landscape as 'made by a group of people who modify the natural environment to survive, to create order, and to produce a just and lasting society'; there are obvious parallels with Sauer's vision of landscape in terms of the material transformation of nature by cultural groups. Further, as Cresswell (2003, p.271) writes, Jackson 'followed the Berkeley line in . . . his distrust of formal theory'. For Jackson, as for Sauer as for Hoskins, landscape exhibits – is – a straightforward materiality and thereness, it is a palpable reality of objects and patterns that the eye can see and the hand can touch. In turn it demands to be investigated and spoken of via a limpid empiricism, concrete, first-hand observation translating into clear, direct, descriptive prose. For Meinig (1979a, p.229), the power of Jackson's vision of landscape lay in 'the naïve fresh look, unordered by orthodoxy; the clean prose, unsullied by jargon; the conversational tone, unstrictured by analytical forms'. In this way his project comes particularly close to that of the Berkeley School, and Jackson did establish direct contact with Sauer and his followers, participating in seminars and giving lecture courses at the University of California in the 1950s and 1960s.

Beyond an empirical, descriptive, first-hand epistemology, a feature of Jackson's work further chiming with that of Sauer and Hoskins is a particularly anti-modern, anti-*state* sensibility. While Jackson embraced and in fact shaped a vision of the emerging everyday features of the post-war south-west American landscape – the dusty highways and neon-lit strips, the scrublands and trailer parks – he did so in part as a rebuke to what he considered to be the imaginative deficiencies and prejudices of both official and academic discourse. Just as Hoskins's bemoaned the centralised state planning which in his view blighted the post-war English landscape, so for Jackson the abstract scientific rationality of American planning, in hock to utopian notions of technologically driven progress, and wedded to inappropriate notions of aesthetic form and proper function inherited from the old-world picturesque landscapes of Europe, worked to render the contemporary American scene invisible, stifling the creative aspirations of the individual and dismissing altogether the backyard beauty of the vernacular. To remedy this situation, Jackson argued, 'we must see the highway not through the eyes of the traffic engineer or economic planner, but through those of ordinary citizens . . . and we must see the garish architecture, shrill signs and insistent lights not as a "longitudinal slum" but as "a kind of folk art"' (Meiniga, 1979, p.217).

### 2.4.3 Landscape from artefact to symbol

In its empirical, artefactual emphases, its anti-state, anti-abstraction ethos, and not least in its individuality and idiosyncrasy, Jackson's vision of landscape clearly has elements in common with those of Hoskins and Sauer. But in something of a contrast, Jackson also repeatedly stresses and explores the role of landscape as a *symbolic* as well as material resource – landscape as a source and repository of myth, imagination, symbolic value and cultural meaning. For Denis Cosgrove (1998, p.xv), Jackson's achievement was precisely to have alerted human geographers to 'those issues of myth, memory and meaning which invade landscape's material existence'. For example, for the very first issue of *Landscape* he penned an essay entitled 'Chihuahua: as we might have been', exploring the differences between the cultural landscapes of Mexico and New Mexico, divided along the Rio Grande by the US–Mexico border. Jackson made the point that while Mexico to the south of the river and New Mexico to the north were topographically and climatically very similar, from the air the two landscapes had a distinctive appearance, each exhibiting evidently different styles and systems of agriculture, settlement and architecture. The obvious point – one that Sauer, for example, draws (see Box 2.1) – was that these differences were a result and reflection of cultural differences between the USA and Mexico; two nations with differing cultural traits, styles of use and occupancy. Jackson makes this point of course. But he also goes further, far beyond the mapping of cultural difference as exhibited in the material form of buildings and field patterns, to write in a much more speculative and discursive fashion about how the visible differences in these two landscapes illuminated differences of belief with regard to social organisation, economic theory and cultural values. As Meinig (1979a, p.211) notes, 'if this was geography, it was a far remove from what was usually found under that name in the classrooms of America. If this was travel literature, it was a far remove from what was usually served the American tourist' (p.211). Jackson, as already noted, had majored in literary and historical studies at Harvard, and this background in the arts and humanities rather than the field sciences, in conjunction with his unique status as a commentator unrestricted by the evidentiary demands placed upon academic publication, perhaps accounts for his ability to shuttle productively between evocative descriptions of material, 'real' landscapes, and landscape considered as a variegated set of values and meanings within Western artistic and literary discourses. He was thus acutely aware that modern ideas of landscape in terms of scenery have their origins in the visual arts of the Renaissance (see Jackson, 1984b; and Chapter 3), and that beyond this, the 'intellectual shaping [of landscape] in America . . . has drawn upon deep resources of myth and memory offered by both Western Classical and Judeo-Christian cultural traditions' (Cosgrove, 1998, p.xi). Just two examples of the interpretative and discursive style of writing and thinking attendant upon this awareness of the intellectual and symbolic dimensions of landscape would be an article on the influence of Newtonian rational and religious cosmology upon conceptions of vastness and order in eighteenth-century America (Jackson, 1984b), and an essay on the symbolic meaning of stones (as fertility objects, as gateways, etc.) (Jackson, 1984b).

In retrospect, it was probably this signature combination of an openness to issues of myth, symbolism and belief in landscape with a commitment to valuing the everyday world of contemporary America that so endeared J.B. Jackson to the North American 'humanistic geographers' of the 1970s. As these were galvanised by a rejection of the positivist, scientistic approach to human geography, and a desire to embrace and elucidate the sometimes deeply felt associations between people and places, in particular with respect to notions of 'home' and belonging', the points of shared interest were clear. The highpoint of this association between Jackson and the geographical humanists – and, it might be argued, of the entire humanistic project in geography – was a collection edited by Donald Meinig (1979b) entitled *The Interpretation of Ordinary Landscapes*. In addition to containing a piece by Jackson, the final section of this volume, called 'Teachers' and written by Meinig himself, consisted of a double intellectual biography of Jackson and Hoskins, cementing their position, but especially Jackson's, as totemic figures within geographies of landscape.

The collectively written introductory essay to *Ordinary Landscapes* further testifies to J.B. Jackson's influence. First, and obviously, it emphasises the 'ordinary', everyday and vernacular qualities of landscape as providing orientation, purchase and justification for human geography, in so far as, 'in its focus upon the vernacular, cultural landscape study is a companion of that form of social history which seeks to understand the lives of ordinary people' (1979b, p.6). Second, however, in a manoeuvre that signals a deci-sive break from an empiricist understanding of landscape in geography, the authors go on to note that ordinary landscapes are suffused with imaginative meaning and collective beliefs and axioms, such that we must regard 'all landscapes as symbolic, as expressions of cultural values, social behaviour' (*ibid.*). This coupling of the vernacular and the symbolic is quintessentially humanistic, and it remains Jackson's most visible legacy – what Denis Cosgrove describes as his 'unique capacity to interpret land-scapes iconographically and intelligently, while remaining true to the everyday experience of landscape' (Cosgrove, 1998, p.xi).

**Box 2.3 J.B. JACKSON – 'AGROPHILIA, OR THE LOVE OF HORIZONTAL SPACES'**

There are landscapes in America separated from each other by hundreds of miles that resemble one another to a bewildering degree. Many American towns, even many American cities, are all but indistinguishable as to layout, morphology, and architecture. The lack of variety in much of our man-made environment is recognised by anyone who has travelled widely in this country. Many deplore it, try to escape it, and because they cannot, suppose that America is altogether lacking the kind of landscape beauty characteristic of older parts of the world.

I have not found this to be the case.

(Jackson, 1984a, p.67)

A short essay on 'horizontal spaces', part of the collection titled *Discovering the Vernacular Landscape*, expresses many of J.B. Jackson's signature themes. In particular, it reveals the authority of his narrative voice (like much of his output, it is easily imagined as a lecture or radio broadcast) and the scope of his intellectual ambitions – to project a distinctively American vision of cultural landscape, and to discover a new aesthetic idiom grounded in an appreciation of the everyday, vernacular American scene.

In the first place, in the very repetitivity of the built environment, there is an aesthetic quality:

they are alike for a good reason: they consciously conform to what is a distinctive American style. Classical is the word for it, I think, and rhythmic repetition (not to say occasional monotony) is a Classical trait, the consequence of devotion to clarity and order. But the style also possesses spaciousness and dignity; that is why I relish the similarity between the villages of New England, the similarity between wheatfields whether in Oklahoma or Oregon, or the stately repetitiousness

continued

of North Dakota shelterbelts . . . the traveller in the United States finds evidence wherever he goes of a national style of spatial organisation. He may not care for it, he may prefer a greater variety of romantic confusion; but he cannot fail to be impressed by it.

Impression evolves into a sort of eagle-eyed attentiveness:

I have found that over-familiarity with the scene has compensations: it teaches sensitivity to change, a sharpened awareness of any deviation from the established style; and if the evidence I have garnered during many years of visiting small towns is of any value, it indicates that the American landscape, that is to say, the manner in which we organise our space, is undergoing a remarkable shift.

(p.67)

Before specifying the general shape of this shift, Jackson gives several localised examples of it: the desertion of the upper stories of the buildings along the Main Street of countless small American towns, the concomitant abandonment of upper floors of warehouses by the railroad – 'only on the ground floor, where a small firm makes handbags, is there activity' (p.68) – and the changing organisation of farming landscapes – 'back of the barn is a complex of new cement block buildings – long and low, with gleaming metal roofs' (p.69).

The paradigmatic example of this shift, however, is

a change of another sort: on the outskirts of the town, in the midst of fields, a housing development – what its promoters call a planned residential community – has recently come into being. . . . The development is laid out along a series of curving roads leading to no particular destination. The houses, painted in bright colours, are still too new to have acquired individuality; they lack gardens and all but the slimmest of trees. Still, the development has a quality of its own: it is an orderly composition of clear-cut, well-defined forms, in no way blending into its natural environment.

(p.69)

American beauty emerges in stark, primary-coloured contrasts of culture and nature, the new houses like spacecraft just landed in the fields. But Jackson also detects a more profound, socio-economic process at work in the scene:

> none of the houses is of more than one storey . . . the multi-storeyed downtown block is abandoned; the multi-storeyed residence is converted into a series of one-storey flats, the multi-storeyed barn with its silos is abandoned in favour of the one-storey cement block structure. . . . Clearly America is showing a preference for the horizontal over the vertical organisation of space.
>
> (p.69)

What at first sight seems an important exception – the enormous increase in high-rise structures in our cities – is, I think, merely another and more complex form of horizontality. The modern multi-storeyed office building differs from the earlier example of the form in being essentially a stack of large, uninterrupted horizontal spaces. . . . The small town contains its own share: the supermarket, the shopping center, the motel, the one-storey consolidated high-school, the one-storey hospital, however commonplace they may now be, are still new and are still being built to replace the old vertical counterpart.

(p.70)

These changes, for Jackson, amplify and confirm his own, distinctive sense of difference – between old and new, Europe and America, between picturesque and elite landscape and an altogether dif-ferent, democratic, mobile and expansive American space:

continued

It is evident that Americans now perceive their environment in a new and as yet undefined manner. It is evident that increased mobility, and even more, an increased experience of uninterrupted speed – whether on the highway or the ski slope, or on the surface of the water – bring with them a sharpened awareness of horizontal space.

It is changes such as these – fragmentary and pragmatic – that we should look for when we explore the American landscape. . . . The study and understanding of landscape metamorphosis can nowhere better be undertaken than in the contemporary United States, but it has to be undertaken in the proper frame of mind; and this is largely a matter of accepting our national landscape for what it is: something very different from the European.

(p.70)

## 2.4.4 Landscape, dwelling and mobility

The story of J.B. Jackson's contribution to landscape studies might easily culminate with humanistic geography's excavation of the symbolic dimensions of vernacular landscape. This development to some extent paves the way for the understanding of landscape as a representation of cultural meaning and power relations that later takes centre-stage within the 'new' cultural geographies of the 1980s and 1990s (see Chapters 3 and 4). As Chapter 3 will detail, however, these new geographies were also in part based upon a rejection of humanist principles. But Jackson's work has been championed more recently again by Cresswell (2003), this time as a pioneering account of landscape as a lived, practised milieu, cognate with current phenomenological work (e.g. Ingold, 2001; Wylie, 2002a; Tilley, 2004).

A phenomenological current, one which would inhabit an 'insider's' perspective by weaving landscape as the ground and context of human being-in-the-world, is most evident in Jackson's repeated insistence that the first object of any study of landscape should be the *house*, or dwelling place. This is not simply because houses are an obvious material manifestation of 'culture' in the landscape, but because for Jackson (1995), in an echo of the anthropologist Mircea Eliade's (1973) concept of polarised sacred space, the act of building and dwelling founds, organises and orients landscape. The notion of landscape, while broader than that of home, is thus anchored in dwelling-activities, in practices of everyday life. As Meinig's (1979a, p.228) overview essay summarises, for Jackson, 'the elementary unit in the landscape is the individual dwelling. . . . Thus in the study of landscape "first comes the house"'.

This groundedness in dwelling is at the root of Jackson's most basic and insistent view that 'far from being spectators of the world we are participants in it' (Jackson, 1997a [1960], p.2), and this, finally, is a philosophical and ethical commitment to a vision of landscape as a shared, lived-in world. Jackson does not delve too far into the intricacies of Martin Heidegger's phenomenological accounts of dwelling and being-in-the-world, nor is his work propelled by a particular political credo or critique. But what Cosgrove (1998, p.xiii) terms a 'contextual and democratic' vision colours all of Jackson's writing, even if this is only vaguely outlined – a vision of landscape as the result of labours 'to produce a just and lasting society' (Jackson, 1995, p.43).

At the same time, as Chapter 5 shall discuss in more depth, Heideggerian notions of dwelling are at least partly compromised by their introspective quality, by their romanticisation of the rural and by their tendency to judge certain ways of life as more authentic than others. A focus on dwelling thus might imply a rather static and backward-looking understanding of landscape. Again, however, Jackson slips the net, this time by virtue of his recognition that mobility and movement were a – perhaps even the – key ingredient of the post-war American landscape. This is the aspect of his work that catches Tim Cresswell's (2003) eye, and even enables him to position Jackson as the 'hero' (p.271) of a story in which an understanding of landscape in terms of embodied movement and practice – a distinctly phenomenological understanding – is retrieved from beneath a vision of landscape as one expression of a panoptic and repressive gaze upon the world, one which paralyses, stifles and constrains. For Cresswell,

> Jackson, more than anyone, reveals the limits and tensions of the landscape idea. On the one hand he carefully described the origins and history of landscape in terms of vision, and on the other he enacted a critique of the term through an emphasis on the everyday routines that produce and reproduce actual living landscapes.
>
> (Cresswell, 2003, p.274)

Movement in particular figures here as an everyday aspect of landscape practice that escapes official jurisdiction, so chiming with Jackson's anti-centralisation, anti-bureaucratic ethos, and opening up a space of relative freedom and autonomy beyond a detached, normative, rule-making gaze. This engagement with issues of movement and mobility in landscape was not confined to a cultural history of the issue – although here again Jackson was an innovator, penning an essay on 'The movable dwelling and how it came to America' (Jackson, 1984b). Typically, and for Cresswell (2003) tellingly, Jackson in *Landscape* wrote about the *experience* of movement in everyday contexts such as driving and motorbiking. The description of such experiences offers a new perspective on landscape:

> the perspective of a mobile observer . . . if the equation that links landscape to vision has frequently erased practice, then J.B. Jackson's mobile view of landscape began to show how vision is a practice. J.B. Jackson's way of looking is so much less reliant on that distanced gaze from above and so much more practised – more embodied.
>
> (Cresswell, 2003, p.275)

Here, Cresswell explicitly claims Jackson for an intellectual lineage concerned to 'think practice', in other words to situate practical, routine human actions as, first, the often-overlooked ground and essence of life, second, the central object of intellectual enquiry and, lastly, the method-ological medium through which research should be conducted. For Cresswell, J.B. Jackson, alongside writers such as Michel de Certeau and Pierre Bourdieu, is primarily notable as a stalwart of studies of 'the everyday and unexceptional' (p.280), as opposed to a scholarship more aloofly focused on the discursive realms of art, literature and philosophy.

These ideas coalesce in one of Jackson's best-known essays for *Landscape*, 'The abstract world of the hot-rodder' (Jackson, 1997a [1957–58]), where the central argument is again all about inhabiting landscape, about experiencing landscape from the *inside*: 'rather than standing back and looking, the hot rodders take part in the landscape' (Cresswell, 2003, p.275). Jackson's own description of this experience easily stands the test of time:

> The view is no longer static . . . the traditional way of seeing and experiencing the world is abandoned; in its stead we become active

participants, the shifting focus of a moving, abstract world; our nerves and muscles are all of them brought into play.

(Jackson, 1997a, p.205)

## 2.5 CONCLUSION

This chapter has sought to provide a useful prelude to the remainder of the book, by outlining the work of three of the most noteworthy figures in landscape studies, and through them, a more general history of approaches to landscape up until the 1980s. It is worth noting that this approach in itself says something about landscape studies as an intellectual field – while a history of, say, human geography or sociology would normally be written as a series of movements and paradigms rather than as the story of a series of biographies, the relative lack of a formal structure to 'landscape studies' as a discipline (for example, in terms of university departments, degree courses, etc.) makes it more necessary to focus upon a few key individuals.

A second aim of the chapter has been to consider the past from the perspective of the present, and, for the most part, to picture the work of Sauer, Hoskins and Jackson through the eyes of contemporary interpreters and critics. In this way, while running the risk of imposing the values of the present upon the past, both historic judgement and contemporary resonance hopefully become more apparent. All three figures do continue to resonate, in different ways. Carl Sauer's 'The morphology of landscape' remains a canonical work, for example. But his legacy is different in the UK and the USA. While still foundational to many US geography degree programmes, the work of the Berkeley School is now commonly presented to students in the UK as something to be transcended; its thought and methods existing as that against which the 'new' cultural geographies of landscape of the late 1980s defined themselves. In partial contrast, J.B. Jackson's legacy would seem to be thriving. Having inspired one generation of humanistic geographers to give themselves over to the study of cultural symbolism in landscape, he also now quite differently provides hooks for those seeking to envision landscape as a milieu of lived, embodied practice (and see Wilson and Groth (2003) for further evidence of Jackson's continuing influence). Lastly, from the perspective of contemporary cultural geography, W.G. Hoskins is today perhaps the least visible of these figures. Where once *The Making of the English Landscape* would have been a staple of many geography degree courses (for instance, up until a dozen years ago, when I read it as an undergraduate student), this is now less the case.

Finally, this chapter can be seen as a baseline for the rest of the book. 'Baseline' is perhaps an appropriate word here, because if one theme unites Sauer, Hoskins and Jackson it is the stress they all laid upon landscape as a *material entity*: landscape as first and foremost a real, palpable, worldly presence, something to touch, observe, walk in. Landscape as the fields, mountains, roads and buildings themselves, and not just as a picture of them, or as their dematerialised symbolic meaning. In turn, this emphasis on the materiality of landscape leads all three to think of landscape as something *external* – not just as something really existing, beyond the imagination, but as 'the great outdoors' in a way; the world beyond the classroom, the library and the study. As a material, external topography, landscape, equally for Sauer, Hoskins and Jackson, demanded to be studied via direct observation, through first-hand *empirical* experience.

Material, external, empirical. Chapters 3 and 4 which follow detail how cultural geographers over the past twenty-odd years have negotiated this triangulation of landscape, moving towards a much greater emphasis upon the discursive and symbolic roles of landscapes in the first instance. But these chapters will also show how issues concerning the *materiality* of landscape in particular have repeatedly returned to haunt cultural geography.

# 3

---

# WAYS OF SEEING

## 3.1 INTRODUCTION

At the start of this book, in talking through the tensions that animate the landscape concept, I noted that landscape can be thought of as both something seen *and* a particular 'way of seeing' the world – both the land *and* the gaze upon it. In either case, however, there is a sense that landscape is quintessentially *visual*. And this focus on vision and the visuality of landscape has underscored much writing both within cultural geography and across the arts and humanities. This chapter explores an argument central to this understanding: the argument that landscape is an expression of 'what is normally claimed to be the dominant, even totally hegemonic, visual model of the modern era; that which we can identify with Renaissance notions of perspective in the visual arts and Cartesian ideas of subjective rationality in philosophy' (Jay, 1992, p.179).

In this quote, and in many of the cultural geographies to be discussed in this chapter, landscape is quite closely identified with *landscape art*, a complex and diverse artistic genre evolving from the fifteenth century to the present day, and associated in particular with the visualisation of relationships between culture and nature. As a system for producing and transmitting meaning through visual symbols and representations, landscape art, alongside cognate arts such as cartography, photography, poetry and literature, is a key medium through which Western and in particular

European cultures have historically understood themselves, and their relations with other cultures and the natural world.

This chapter outlines some of the key ways in which cultural geographers have apprehended and used this insight. In discussing *landscape as a way of seeing*, it also speaks more widely about some of the characteristics of what were known as the 'new cultural geographies' of the late 1980s and early 1990s. It should be noted that, in this way, the focus of the chapter is upon the *principles* which have underpinned understandings of landscape in this period, rather than discussing specific examples of landscape analysis. The chapter thus identifies and details three significant metaphors through which the landscape way of seeing has been examined by geographers: landscape as veil, landscape as text, landscape as gaze. Having outlined these three metaphors, the chapter then seeks to identify some of the epistemological principles they share, and thus concludes with a broader discussion of the practice and process of interpreting landscape within new cultural geographies. To begin with, however, it is important to discuss the origins of the landscape way of seeing.

## 3.2 LANDSCAPE AND LINEAR PERSPECTIVE: ART, GEOMETRY, OPTICS

### 3.2.1 Perspective

When – if – we stand in an art gallery and look at a landscape painting, we commonly have an impression of *visual depth*, as if we were looking out through a window onto the fields, the mountains, or the city being depicted. Landscape art *is* this spatial depth, this vision of things – fields, trees, people, buildings – receding and diminishing into the distance. Moreover, what we see in a work of landscape art often appears to be visually realistic, convincing and in proportion; it often appears to approximate our 'natural' or everyday perception of the world. But then we know as well that this is something of a deception. Clearly the canvas is only a flat, two-dimensional surface. The impression of depth, of three-dimensionality, is an illusion.

Linear perspective is the name usually given to the geometric system which enables this illusion – the depiction of three-dimensional spaces and objects upon two-dimensional surfaces: canvasses, walls, screens, photographs. Today, the visual arts are usually placed at the opposite end of the disciplinary spectrum to subjects like mathematics and geometry, but nonetheless it is the case that, however purely aesthetic or decorative some examples may seem, landscape art has its underlying basis in geometric laws of perspective and proportion. Linear perspective, which organises the entire space of a picture around a vanishing point on the horizon, is the underlying geometric spine that allows the hills, fields, buildings and people depicted to stand upright, fully fleshed and visually rendered in a plausible, coherent and convincing fashion. And beyond this technical element, it is difficult to overstate the influence that perspectival techniques of picturing have had more broadly upon both our sense of ourselves and our perception of the world. The language, attitudes and implications of this particular way of seeing might be said to be hard-wired into Western cultures. Thus, as regards the general cultural significance of linear perspective, the art historian and visual theorist James Elkins (1994) feels able to say it is '"our" perspective . . . the one that describes how we view the world, and constitute ourselves as viewing subjects'.

The exact origins of perspectival techniques of visual representation in the West remain a subject of debate (see for example Alpers, 1983; Damisch, 1994; Elkins, 1994; Kemp, 1990). There is, however, general agreement that perspective first began to be used and explored in northern Italy in the mid-fifteenth century, in the city of Florence; in the midst of a culture in the process of rediscovering the values, knowledges and insights of classical civilisations; and also a culture, it has been argued, peculiarly attuned to gauging and measuring space by eye (Baxandall, 1988 [1972]). A notable figure in this culture was the artist and architect Filippo Brunelleschi, designer of the dome of the cathedral of Santa Maria del Fiore, a feature which still dominates the Florentine skyline. Brunnelleschi is further credited by many sources as the inventor (or discoverer) of linear perspective – and hence, by extension, of landscape as a particular way of seeing the world – although the earliest extant and by some way most influential discussion of perspectival theory and practice is Alberti's *De Pictura*, a text which has served for centuries as a training manual for landscape artists.

Samuel Edgerton's (1975) noted account of what he terms the 'Renaissance rediscovery' of perspective centres upon a re-creation of Brunelleschi's (perhaps apocryphal) experimental demonstration of linear perspective to the Florentine populace. In this re-creation, the artist and architect is depicted as offering to the audience a comparison between an *actual* church – the Bapistry of Santo Giovanni – as seen from a particular vantage point, and an *image* he had drawn of it. Brunelleschi's image, an etching subsequently lost, was drawn according to the rules of linear perspective, rules which at this stage he alone knew. For Edgerton, the key to understanding this episode lay in the amazement voiced by the audience on seeing this original perspectival image. They were amazed precisely because Brunelleschi's drawing resembled the real thing so closely. Brunelleschi seemed to have found a method for accurately and realistically representing the world-as-perceived upon a flat surface. Putting himself in the place of a Florentine offered the comparison between image and reality, Edgerton writes '"Oh maestro", one can almost hear the comment, "truly, I see no difference between your painting and our own Santo Giovanni!"' (*ibid.*, p.152). And this is in many ways the cardinal significance of perspective, and of the landscape art it enabled: it was understood to be a realistic, truthful and authoritative representation of space. Perspective enabled a commanding, objective and controlled grasp of space and spatial relations.

## 3.2.2  Perspective, landscape and property

In a seminal paper, 'Prospect, perspective and the evolution of the landscape idea', cultural geographer Denis Cosgrove (1985, p.45) explores the implications for our understanding of landscape of the fact that 'the basic theory and technique of the landscape way of seeing is linear perspective'. Cosgrove focuses upon the socio-historical implications of perspective, and, by extension, landscape, as a 'way of seeing' the world. If by the term 'landscape' we refer to visualisations constructed on the basis of perspective, then from the start, Cosgrove argues, landscape is suffused by connotations of authority, control and ownership. Noting that perspective 'was regarded as the discovery of the inherent properties of space itself' (1985, p.51), he suggests that by implication,

> Landscape is thus a way of seeing, a composition and structuring of the world so that it may be appropriated by a detached individual spectator to whom an illusion of order and control is offered through the composition of space according to the certainties of geometry.
>
> (Cosgrove, 1985, p.55)

Writing from a critical materialist point of view, Cosgrove's concern is to establish connections between the 'authority' of perspective as a scientific vision, its production of, and reliance upon, a detached envisioning subject, and the material transformation of land into private property. Here, perspective, as a geometrical system of perception and representation, 'gives the eye absolute mastery over space' (ibid., p.48). In the same movement, the gaze of the beholder is removed, distanced from the objects and relations which form the field of vision, such that 'perspective directs the external world toward the individual located outside that space' (ibid.). Stressing the overall relation between this system and fifteenth-century artistic humanism, Cosgrove argues that 'linear perspective provides both the certainty of our reproductions of nature in art and underlies the power and authority, the divine creativity of the artist' (ibid., p.52). As a way of seeing, therefore, landscape is the accomplice and expression of a classical subject–object epistemological model, one whose central supposition posits a pre-given reality which an independent subject contemplates, represents and masters from a position of cohered detachment.

However, in a more vital sense, Cosgrove's paper is an attempt to underscore the relations between perspectival vision and the materiality of property relations within capitalist and mercantile economies. In the plainest possible sense, if, through perspective, 'visually, space is rendered the property of the individual detached observer' (ibid., p.49) it thus follows, for Cosgrove, that the landscape way of seeing involves 'control and domination over space as an absolute, objective entity; its transformation into the property of the individual or the state' (ibid., p.46). Landscape is thus a 'way of seeing' complicit with the needs and purposes of an elite, patrician group of capitalist property owners. In a generic sense, therefore,

> One of the consistent purposes of landscape painting has been to present an image of order and proportioned control . . . there is an inherent conservatism in the landscape idea, in its celebration of property and of an unchanging status quo, in its suppression of tensions between groups *in* the landscape.
>
> (Cosgrove, 1985, p.58, original emphasis)

Although Cosgrove does not, in this paper, dwell at any length upon the precise ways in which this thesis is borne out, for example by the evidence of eighteenth- and nineteenth-century landscape art, its implications are nevertheless clear. Not only does perspective establish vision as the *privileged sense* in epistemological terms; the landscape 'way of seeing' which develops from perspective further becomes in itself the *sense of the privileged*.

---

**Box 3.1  DENIS COSGROVE – *SOCIAL FORMATION AND SYMBOLIC LANDSCAPE***

'The argument here is that the landscape idea represents a way of seeing – a way in which some Europeans have represented to themselves and to others the world about them and their relationships with it, and through which they have commented on social relations' (Cosgrove, 1998 [1984], p.1).

*Social Formation and Symbolic Landscape*, by Denis Cosgrove, was a path-breaking text for cultural geographers in the 1980s and 1990s. In contrast to most texts on landscape by geographers hitherto, it focused squarely upon 'the idea of landscape, its origins and development as a cultural concept in the West since the Renaissance' (p.1). And it also gave consistent expression to a cultural Marxist understanding of the relationships between landscape art, vision, and social and economic relations, especially as these were expressed through patterns of property, commerce and landownership. In this vein, Cosgrove writes: 'The landscape idea emerged as a dimension of European elite consciousness at an identifiable period in the evolution of European societies: it was refined and elaborated over a long period during which it expressed and supported a range of political, social and moral assumptions and became a significant aspect of taste' (p.8).

The 'identifiable period' referred to here is that span of the fifteenth and sixteenth centuries commonly called the Renaissance. Here also, 'European elite consciousness' is understood to have northern Italy as its centre. Cosgrove is especially interested in Renaissance Italy as the cradle of the landscape idea, and two substantial chapters of *Social Formation and Symbolic Landscape* are devoted to examining this context.

The second of these chapters focuses upon sixteenth-century Venice:

Venice was in certain key respects unique as an Italian city state, particularly in its relationship with land and agriculture, its economic development represents the greatest achievement of Italian mercantile capitalism in early modern Europe, and the achievements of its painters, architects and urbanists one of the most enduring contributions to the European landscape idea. In the early sixteenth century Venice itself was redesigned as a consciously symbolic landscape while on the *terraferma* rural landscapes were created which would provide a vocabulary later in those lands which succeeded to Venetian commercial prominence: England and the USA.

(p.103)

The *terraferma* was a swath of agricultural land stretching westward from Venice. As this gradually became the demesne of the merchants of Venice in the sixteenth century, and thus the villa-strewn setting for a languid if not degenerate aristocratic life, so it also began to be painted and poeticised by a group of artists whose names would remain renowned centuries later: Giovanni Bellini, Giorgione, Titian and Paolo Veronese: 'They were the first to capture the mood of literary arcadia in paint, and their work is the central source of a tradition of European landscape painting stretching through Claude Lorrain to English eighteenth-century and American nineteenth-century ways of perceiving and painting landscape' (p.122).

As Cosgrove notes, this Venetian landscape school introduced and popularised a number of recurrent motifs. Whereas the art of Florence, the Italian city-state in which the rules of linear perspective had first been uncovered and applied in the fifteenth century (see Section 3.2), was characterised by 'an interest in the formal structure of the painting, the accurate rendering of realist space', Venetian art displayed 'a greater emphasis on skill in handling painterly materials: the various media, brushwork, finish, and the controlled use of colour and light' (pp.103–104). In terms of landscapes set in an idealised version of the *terraferma*, these

continued

techniques produced a soft, luminous, pastoral vision, a rural idyll, narrated through reference to episodes from Classical mythology.

> The Renaissance reverence for Antiquity, together with the increasingly aristocratic character of Italian ruling-class culture, ensured that the landscape idea was given a strongly theoretical and classical foundation and reference. From Horace and Virgil came ideas of a Golden Age of harmony between a leisured human life and a willingly productive nature, or pastoral youth and innocence in a bucolic woodland glade, and of the smiling landscape of holy agriculture as an emblem of a morally and socially well-ordered estate.
>
> (p.142)

The crux of Cosgrove's argument, however, is that this 'idyllised' vision of landscape covered up and concealed the actual material conditions of Italian country life:

> A view at ground level allows large areas of sky to provide an intense source of light bathing the picture. Human figures are painted full-face or in sharp profile, at rest and adding to the serenity of the landscape. The landscape is constructed in a series of low undulations and rounded hill forms differently highlighted or in shadow. In the background runs the lines of the Alps. . . . [However] this is far from the landscape of daily life, but an imaginative harmony in which no vulgar *mezzardi* or *livelli* struggle to meet their rents nor where arduous work upsets the balanced calm.
>
> (pp.123–125)

Thus, landscape for Cosgrove is from the first a way of seeing that chimes with elite and aristocratic visions of human society and nature. Such visions are often profoundly distant from the actuality of working and living in landscape, and should be understood as imposing an aesthetic and moral order from afar. Art such as that of sixteenth-century Venice,

> parades the love of landscape as the setting for a refined
> and other-worldly Arcadian life . . . the idea of landscape
> in Renaissance Italy can best be understood in its alignment
> with the shifts of inflections of power and control in society.
> Whether objective or idealist in basis, urban or rural in reali-
> sation, the landscape idea remained above all an appropriation
> of the visual scene by sense and intellect rather than an active
> engagement with it in the processes of organic and productive
> human life.
>
> (pp.140–141)

## 3.3 CULTURAL MARXISM, ART HISTORY AND LANDSCAPE

The language which Cosgrove uses in the presentation of his argument
serves to indicate a particular critical allegiance. In describing landscape
as a 'visual ideology' (1985, p.47), in alluding to the manner in which art
and literature serve to 'sustain mystification' (ibid., p.58) and above all in
speaking of landscape as a 'way of seeing', his paper positions itself within
a cultural Marxist interpretative tradition most notably associated with the
work of Raymond Williams and John Berger, and is concerned to present
a critical account of the complicity of artistic and literary genres with
evolving capitalist systems of production and property ownership. *Ways of
Seeing* is the title that Berger (1972) gave to his critique of the conservative
art history of the time, especially that exemplified by Clark (1969).

A quote from the critic Walter Benjamin, to whom Berger acknowledges
a considerable debt, establishes one principle tenet of the cultural Marxist
critique of landscape art: 'The manner in which human sense perception
is organised, the medium in which it is accomplished, is determined not
only by nature but by historical circumstances as well' (1992, p.216). Thus
the task of a critical art history, as expressed by Berger, involves a critical
recognition of the historicity of perception, and of art. Such a recognition
enables a purposeful alteration of interpretative context, from an aesthetics
to a politics of vision, from a landscape judged through reference to 'internal' criteria ('beauty, truth, genius etc.' – Berger, 1972, p.5), to one whose
varied 'meanings' are accessible through an examination of the historical
conditions of its production. In Berger's polemical account, the situation

necessitating such a critical shift is explicitly stated: 'The art of the past is being mystified because a privileged minority is striving to invent a history which can retrospectively justify the role of the ruling classes' (ibid., p.5). For Berger, such 'mystification', in the form of a dehistoricised and supposedly universal visual analytic, is thus a scaffolding deliberately erected to evade 'the only confrontation that matters' (ibid., p.9), the question of socio-economic inequality, of which a particular painting may be said to be an expression.

In the particular art-historical arena of English landscape, Barrell (1980) and Bermingham (1986) provide lucid exemplars of a Marxist-inflected interpretative stance. Barrell especially, focusing upon the depiction of the 'poor' in rural landscapes by Constable, Gainsborough and Morland, is clear as to what he believes to be the ideological function of such works, and also as regards the interpretative potential they nevertheless continue to offer:

> The art of rural life offers us the image of a stable, unified, almost egalitarian society; [thus] my concern in this book is to suggest that it is possible to look beneath the surface, and to discover there evidence of the very conflict it seems to deny.
>
> (Barrell, 1980, p.5)

Examining 'the constraints – often apparently aesthetic, but in fact moral and social – that determined how the poor could, or rather how they could not be represented' (ibid., p.1) he suggests that *industriousness* was a key motif in the production of an acceptable symbolic idiom. This forms part of an implicit criticism of previous accounts of the period, which interpreted the increasing number of depictions of everyday rural scenes in terms of a demand for greater 'realism' on the part of a buying public tired of the mythical and allegorical themes dominant in the seventeenth-century landscapes of Claude Lorrain and Nicolas Poussin. By contrast, Barrell argues that scenes of 'labourers at work' fulfilled both a morally prescriptive desire to condemn 'idleness' and, at the same time, sought to *naturalise* the workers through identifying them with supposedly 'timeless' rural activities. Thus the workers were transformed into 'distant, generalised objects of fear and benevolence' (ibid., p.3), and thus they could become 'an acceptable part of the décor of the drawing rooms of the polite' (ibid., p.5). For such a representation, Barrell argues, sought 'to offer a reassurance that the poor of England were, or were capable of being, as happy as the swains of Arcadia' (ibid., p.6). And in this way the 'realism' associated with the paintings of this era may be seen as double-edged. Its purpose, for Barrell, was 'at once to reveal more and more of the actuality of the life of the poor, and to find more effective ways of concealing that actuality' (ibid., p.16).

In her study of almost precisely the same period, which focuses more specifically upon the connections between 'rustic' landscape depiction and the enclosure process, Bermingham (1986) draws upon a similar analytic. Noting that the 'objectivity' of traditional art history is 'a myth that enforces the very cultural values it presumes to stand outside of' (p.3), her thesis proposes that

> There is an ideology of landscape, and that in the eighteenth and nineteenth centuries a class view of landscape embodied a set of socially, and, finally, economically determined values to which the painted image gave cultural expression.
>
> (Bermingham, 1986, p.3)

However, like Barrell, Bermingham's subsequent analysis of the ideological function of landscape is propelled by the critical assumption that 'art' as such is not straightforwardly a 'reflection' or 'outworking' of an underlying economic order but is rather dynamically involved in the production and sustenance of particular ideologies. Again, therefore, this offers the possibility of critical purchase, 'For though art produces ideology and exists within it, it also registers the inconsistencies within ideologies and pinpoints the places where their totalising worldview threatens to unravel' (ibid., p.4).

## 3.4 CULTURAL MARXISM AND CULTURAL GEOGRAPHY: LANDSCAPE AS 'VEIL'

### 3.4.1 The emergence of new cultural geographies

The quote above, along with its associated analytical rubric, finds an echo in geography through Cosgrove's (1983, p.10) early call for a 'radical cultural geography' based upon 'an aesthetic grounded in the recognition of how landscapes sustain and elaborate the symbolic code of bourgeois society'. Yet during the early 1980s, such an analysis of landscape subsisted in a subterranean fashion within cultural and historical geographies at that time more guided by traditional humanistic scholarship, for example in articles by Cosgrove and Thornes (1981) on John Ruskin and Daniels (1982) on Humphrey Repton, which appear in collections more explicitly concerned with the 'subjective meaning' of landscapes (respectively, Pocock, 1981; Gold and Burgess, 1982). This intricate process of emergence may be at least partially explained in relation to a broader debate in which both radical and humanistic geographies sought to critically oppose a prevailing positivist paradigm. As Cosgrove (1985) notes in the paper previously discussed, and as was detailed in the previous chapter, in this debate 'landscape' was a term mostly associated with a humanistic per-spective, because of both its disciplinary alignment with a descriptive, Sauerian cultural geography and latterly its more diffuse evocation of vernacular symbolic meanings (Meinig, 1979a).

In the early to mid-1980s tensions thus emerged in human geography around the question of the interpretative use of the term 'landscape', tensions most clearly evidenced in a paper by Stephen Daniels (1985). Entitled 'Arguments for a humanistic geography', Daniels's paper is a sustained critique of the humanistic endeavour. Its central point is that a 'historical myopia . . . afflicts much humanistic geography' (p.145). This short-sightedness, even blindness, Daniels argues, is evident at all scales. Conceptually, humanistic geographers are seemingly content to pursue the 'emotional' in favour of the 'rational, the 'subjective' instead of the 'objective', in a facile way which assumes the 'purity' and 'authenticity' of individual experience, and which is reliant upon a misguided and partial appropriation of the phenomenological search for 'essences'. This in turn impacts, in particular, upon their studies of landscape in art and literature (e.g. Pocock, 1981), which, for Daniels, are instances of mys-tification that elevate the artist's creativity to the status of an ineffable 'mystery', and transform the work of art into 'a vehicle for transcendent truths' (1985 p.149). Ultimately then, humanistic geographies of land-scape are thus themselves complicit in the ideological separation of the personal from the public, the social and the economic which has been central to a bourgeois conception of the world since the early nineteenth century.

Alongside the work of Cosgrove (1985), Daniels's paper helps to signal a sea change of sorts in the meaning of 'landscape' within cultural geog-raphy. No longer understood as a purely aesthetic product, representing (in art) the imaginative powers of the artist, and (in life) the creative possibilities and pregnancies of human being in general, landscape, as an *artistic genre*, is instead enrolled within historical and material processes. Most precisely it functions symbolically on behalf of an elite, and its visual aesthetics at once express and occlude particular socio-economic relations. This change is expressed and summarised most forcefully in a subsequent influential paper by Daniels: 'Marxism, culture and the duplicity of landscape' (1989). Here, through a detailed exploration of the intellectual careers of Raymond Williams and John Berger, a genealogy of relations between Marxism and 'culture' is outlined, from William Morris, John Ruskin and E.P. Thompson through to what Daniels calls the 'postmodern marxism' of Debord, Barthes and Baudrillard. The narrative thread which links this assortment of theorists is the concept of landscape's *duplicity*. Citing as an instance, 'the complicity of the realist tradition of western painting with capitalist strategies of economic, social and sexual appropriation and control' (p.213), Daniels argues that landscape, as a way of seeing, is duplicitous, because whilst on the one hand it offers a redemptive, transcendent and aesthetic vision of sensual unity with nature, on the other it operates as a smokescreen concealing the underlying truth of material conditions and manipulating our vision such that we have become unaware of the distancing which separates us from the natural world.

At this point it is important to note that the various writings of Cosgrove and Daniels in the 1980s, in particular their edited collection *The Iconography of Landscape* (1988), have exercised an influence beyond the study of 'landscape' per se, in so far as they were instrumental in the emergence and formation of what was once known as 'new' cultural geography. Indeed, the opening line of Daniels's (1989, p.196) paper runs: 'In an area of overlap between radical and humanistic geography a new cultural geography is emerging.' Equally, the rubric of this 'cultural turn' is framed explicitly by the materialist analysis of landscape in a paper by Cosgrove and Jackson (1987), wherein culture, as a 'signifying system', presents, 'A deceptive appearance of naturalness and transparence, concealing an opaque, distorting, arbitrary mechanism of representation, a process of ideological mystification' (p.100). In retrospect, however, the apogee of the cultural Marxist critique of landscape, along with the roots of its eventual displacement, is perhaps best represented in Cosgrove and Daniels's editorial introduction to *The Iconography of Landscape*. At the beginning the editors offer the following definition:

> A landscape is a cultural image, a pictorial way of representing, structuring or symbolising surroundings . . . a landscape park is more palpable but no more real, no less imaginary, than a landscape painting or a poem. . . . And of course, every study of a landscape further transforms its meaning, depositing yet another layer of cultural representation.
>
> (Cosgrove and Daniels, 1988, p.1)

### 3.4.2 Landscape: representation, the visual and the veil

From this definition, it is apparent that there are three central aspects to the cultural Marxist conception of landscape. First, and most obviously, landscape is understood as being *always already* a representation. Its ambit is understood to be the symbolic, the significant, the pictorial, the imagistic. This indexes on one level a shift from a conception of landscape as a material record of (to borrow a phrase) 'man's role in shaping the face of the earth', towards a more decidedly interpretative and discursive understanding, wherein landscapes are paintings, poems, prose: products of cultural imagination.

Second, landscape is intensely, essentially, *visual*. More precisely, as Duncan (1995, p.414) notes, it is '*painterly*'. Instead of involving the empirical description of a delimited area by a researcher, as in the traditional Sauerian prescription, the study of landscape is held to involve the critical interrogation of paintings and images. Within this basic distinction, moreover, a more fundamental theoretical shift is registered. Every 'landscape', as a visual space produced by perspective, is a *particular, fixed, point of view*. In other words, it is no longer a synthetic term with 'regional' associations. A *succession* of perspectives from various points in the 'field' no longer 'build up' into an 'overall picture' of the landscape. Rather, in the movement from chorology to critique a *single gaze* is identified, the vision of the individual subject. The landscape way of seeing, moreover, in a sense creates and sustains the subject, through its establishment of, 'A fixed relationship between object and subject, locating the viewer outside of the picture and outside of the relations being depicted' (Thomas, 1994, p.21).

As Raymond Williams (1985, p.26) famously argued, 'a working country is hardly ever a landscape. The very idea of landscape implies separation and observation'. Within the network of relations established by the landscape way of seeing, therefore, *the subject* becomes the nexus of vision; the 'receptor' of the depicted objects, the 'processor' of visual information and, crucially, the 'projector' of a specific 'gaze'. Thus placed at the centre of analysis, the presence of this subject collapses any potential distinction between envisioning and symbolising; as Cosgrove and Jackson (1987, p.99) remark, 'landscape is the history of a way of seeing, or, better, of representing'.

Finally, the materialist position offers a quite specific interpretation of the purposes and status of visual representations of landscape. In stating that 'a landscape park . . . is no more real, no less imaginary, than a landscape painting or poem' Cosgrove and Daniels seemingly collapse an epistemological distinction between 'image' and 'reality' (even 'culture' and 'nature'). But a distinction between the 'image', or the surface appearance, and the 'underlying truth' is in fact the entire basis of the materialist analysis, which presupposes that all representation has an ideological function. To adopt Williams's terms, this function involves the elision of the distinction between the *real* conditions of a 'working country' and their *fictive* depiction in 'landscape' form. The cultural Marxist critique of landscape, already a critique of a particular subject position, is, by extension, a critique of the 'perspective' of a particular socio-economic group. The landscape way of seeing is thus understood to be the preserve of an elite, it symbolises their dominion over the land in the very act of 'naturalising' it, of making its particular representation seem the natural order of things.

Thus, as Matless (1992, p.41), notes, for cultural Marxists the idiom through which landscape is to be approached, interpreted and categorised, primarily involves the metaphor of the *veil*. More precisely yet, Bender (1994, p.245) speaks of landscape as being a 'proprietorial palimpsest'. Landscape, as a particular type of visual representation, mystifies, renders opaque, distorts, hides, occludes reality. In John Berger's (1972, p.41) own words (the ones also used by Denis Cosgrove (1998 [1984]) in concluding *Social Formation and Symbolic Landscape*): 'Sometimes a landscape seems to be less a setting for the life of its inhabitants than a curtain behind which their struggles, achievements and accidents take place.'

## 3.5   LANDSCAPE AS TEXT: SEMIOTICS AND THE CONSTRUCTION OF CULTURAL MEANING

### 3.5.1   Introducing landscape as text

As Chapter 4 will discuss, the notion that landscape can be understood as an ideological veil has proved enduringly productive for cultural geographers working in many different contexts. But, at the end of their introduction to *The Iconography of Landscape*, Cosgrove and Daniels (1988) themselves begin to challenge the axioms of cultural Marxism, in particular the 'verticality' of an analytic which seeks to twitch aside the 'veil' of surface appearance, and so penetrate to the 'reality' concealed beneath its folds. They note that 'every culture weaves its world out of image and symbol' (p.8). Of John Berger's and Raymond Williams's writing, they say, 'we recognise these Englands for what they are: images, further glosses' (*ibid.*). And they conclude with the following comment:

> From a post-modern perspective landscape seems less like a palimpsest whose 'real' or 'authentic' meanings can somehow be recovered with the correct techniques, theories, or ideologies, than a flickering text displayed on the word-processor screen whose meaning can be created, extended, altered, elaborated and finally obliterated by the merest touch of a button.
>
> (Cosgrove and Daniels, 1988, p.8)

Herein, there may be discerned the contours of a new interpretative context for landscape, one which began to emerge in cultural geography through the later 1980s and early 1990s. A transposition of meaning occurs. The task of the landscape critic is no longer to rent the veil asunder, but to search amidst its folds, along the 'weave' of its 'fabric'. To pursue this metaphor even further, what has now become interesting about the veil is no longer its function, but its *texture*. In this section, consequently, I will discuss the idea that landscape may be understood, and examined, via the metaphor of text.

To liken landscape to a written text – to say that a landscape is like a book – is to open up a series of interpretative avenues. For example it invites questions regarding authorship and interpretation – writing and reading. Who is it that has written the landscape? Which individuals or groups are its principal authors? What is the narrative of the landscape, what story does it tell? Does the landscape have just one plot, or is it composed of many overlapping and even competing storylines? Equally, the landscape-text obviously requires a readership, an audience, in order to come alive. But how will the landscape be read? Is it written in a language that we understand? Or will we need to learn new languages and develop new techniques for reading and interpreting the landscape, if we wish to understand it more deeply?

A last question is most telling: what can cultural geographers, with their critical agendas and affiliations with intellectual movements across the humanities, bring in particular to the task of reading landscape-texts? This question is worth asking, because the notion that landscapes are like books is one that has a pedigree stretching back and across through the more empirical and field-based disciplines of landscape archaeology and landscape history. For example, Muir's (2000 [1981]) guide to the practice of fieldwork in landscape history is titled *Reading the Landscape*, where 'reading' refers largely to knowledgeable field observations, and where the landscape is a book in the broadest sense of being a history of those who have lived and died in it.

Through the late 1980s and early 1990s, however, some new cultural geographers began to turn towards the interpretative techniques of literary and cultural theory, and hence to conceptualise landscape as text in a new and distinctive fashion. Instead of being a historical record, the landscape-text – here composed of both the material landscape 'itself', *and* its representation in art, maps, texts and other imagery – is understood as being organised around questions of power and authority. In consequence, the task of the critical reader centres upon uncovering the hidden codes and meanings, and unquestioned assumptions, which in actuality structure how the text of landscape is read.

This distinctively 'structuralist' concept of landscape-as-text is expressed in a noted paper on '(Re)reading the landscape', by the then US-based cultural geographers James and Nancy Duncan (1988). The terms they use also indicate the many similarities between their agenda and that of writers such as Cosgrove and Daniels:

> It can be argued that one of the most important roles that landscape plays in the social process is ideological, supporting a set of ideas and values, unquestioned assumptions about the way a society is or

should be organised. . . . If landscapes are texts which are read, interpreted according to an ingrained cultural framework of interpretation,
if they are often read 'inattentively' at a practical or nondiscursive
level, then they may be inculcating their readers with a set of notions
about how the society is organised: and the readers may be largely
unaware of this.

(Duncan and Duncan, 1988, p.123)

In describing landscape as a 'veil', geographers such as Cosgrove and Daniels
were concerned to stress how representations of landscape in art per-
petuated and enshrined the world view of certain elite groups. Equally, in
describing landscapes as 'texts', Duncan and Duncan are here laying stress
upon the manner in which landscapes may be understood as expressions
of cultural power. The text of landscape conveys and cements certain
ideological narratives about the organisation of society and relationships
between culture and nature. 'Reading', therefore, is not an innocent, free or
whimsical activity. Instead the metaphor of landscape-as-text calls attention
to ways in which particular, dominant readings are expressed and reproduced by powerful cultural elites. The critical agenda devolving from this
realisation is one in which cultural geographers are charged with exposing
and laying bare the mechanisms through which dominant ideas and beliefs
are reproduced via landscapes, and with supplying alternative readings from
a theoretically informed perspective.

### 3.5.2  Landscape, text and intertextuality

This textual understanding of landscape, as elaborated in a number of books
and papers by Duncan and Duncan (1988, 1992; Duncan, 1990; see also
Barnes and Duncan, 1992; Duncan and Ley, 1993), owes much of its
derivation to the work of the French literary critic Roland Barthes and his
elaboration of the interpretative methods of semiotics. In fact, Barthes
intellectual career to an extent mirrors the conceptual development of new
cultural geographies. Throughout the late 1950s and 1960s, he undertook
analyses of cultural texts (films, books, adverts), in some ways quite similar
to those supplied by the cultural Marxism of Berger and Williams. Semiotics
is the study of the production and communication of cultural meanings
by signs – by systems of signifiers such as the mass media, advertising, the
film industry, the popular press and so on. The study of cultural signs, as

developed by Barthes, aimed to 'account in detail for the mystification which transforms petit-bourgeois culture into a universal nature' (Barthes, 1990, p.9). This process of ideological mystification Barthes termed *myth*: myth thus 'has the task of making an historical intention a natural justification, and making contingency appear eternal' (*ibid.*, p.142). In *Mythologies* (1990, [1957]), one of his most early and still most-noted works, Barthes illustrated this process through a series of short, pithy essays often focused on advertising and consumption, 'analysing the signifying systems of fashion, striptease, Racinian tragedy and steak and chips with effortless brio' (Eagleton, 1983, p.135). To reiterate, throughout this work, Barthes's aim was to show how seemingly unremarkable cultural processes (such as the the iconic place of steak and chips in French culinary culture and family eating) in actuality worked as systems of codes and signs through which dominant bourgeois cultural and moral norms were sustained. This sort of analysis, in which the surface of a cultural text is probed so as to reveal the operation of underlying systems and structures, is commonly termed 'structuralist'.

Barthes's later work, however, is characterised by a gradual move away from the interpretation of texts and images in terms of ideology and myth, towards what can be called a more 'poststructuralist' style of textual analysis. In his most celebrated essay, 'The death of the author' (Barthes, 1977, p.146), for instance, he attacks the privileging of the author over the reader in traditional literary criticism. Textual interpretation had traditionally focused upon single texts (books, plays, poems) in isolation from the wider cultural fields in which they were produced, and had commonly paid much attention to authorial character, style and motivation. Barthes offers instead the more anonymous idea of *intertextuality*, arguing that 'the text is a tissue of quotations drawn from the innumerable centres of a culture'. 'Text' in this expanded sense thus refers to the entire discursive ensemble of words, signs, images and so on. And crucially, this is a field in which the critic is always already situated, such that his or her 'reading' activities are simply part of an ongoing, limitless, intertextual production. In this way, as Barthes puts it, the Text is 'woven entirely with citations, references, echoes, cultural languages, antecedent or contemporary, which cut across it through and through in a vast stereophany' (*ibid.*, p.160).

Barthes's theorisations of the Text constitute a direct attack upon a realist or commonsensical understanding of language and writing, in which it is assumed that meaning and understanding are produced through a correspondence between words (spoken or on the page) and the 'real' things to which they refer. Instead, in the poststructural formulation 'meaning' is located, or, more accurately, *constructed*, within a referential, discursive and intertextual realm.

Notions of the cultural construction of meaning have become common parlance across human geography. Those studies of landscape which are cognate with this theoretical literature through their use of the metaphor of landscape-as-text have, however, been in some ways quite selective in their own reading of authors such as Barthes. Duncan and Duncan (1988, p.120), for example, appear on one level to concur with the principle tenets of poststructural textual analysis, noting approvingly that it:

> Demystifies the illusion of texts as unified original creations of a Cartesian subject. They deny the authority of the author. In rejecting the view that texts are referential, they also reject the idea that texts are representations or reconstructions of the real world. These descriptions suit landscapes well because landscapes are usually anonymously authored; although they can be symbolic, they are not obviously referential and they are highly intertextual creations of the reader.

Equally, they suggest that, 'Texts have a web-like complexity, characterised by a ceaseless play of infinitely unstable meanings. This picture is interesting, not only from a literary standpoint, but also because it resembles landscapes in many respects' (*ibid*., p.118). There is something of an ambiguity here regarding the literal and/or metaphorical import of the 'landscape-as-text' trope. But Duncan and Duncan also note the avowedly spatial idiom in which poststructural 'textuality' is often cast. Text is understood to be 'a space in which the reader as writer can wander, in which signifiers play, signifieds become signifiers in an endless process of deferment' (*ibid*.). And so, conversely, it is possible to state that *landscape* is 'an ordered assemblage of objects, a text, it acts as a signifying system through which a social system is communicated, reproduced, experienced and explored' (Duncan, 1990, p.17).

However, throughout their own assemblage of texts on the subject, Duncan and Duncan (1988, 1992; Duncan, 1990), whilst endorsing a Barthean conception of myth and ideology, remain ambivalent as regards the epistemological implications of the fully fledged poststructural reading of textuality. For example, they comment that, 'Through Barthes' eyes one sees the world exposed and demystified, one's "natural attitude" towards the environment is shattered as the apparent innocence of landscapes is shown to have profound ideological implications' (Duncan and Duncan, 1992, p.18). Thus, a 'vertical' analytic similar to that of cultural Marxism is reinstated, such that 'meanings are always buried beneath layer of ideological sentiment' (Duncan and Duncan, 1988, p.117). Ideology and landscape become once again symbiotic, in the sense that, 'by becoming part of the everyday, the taken-for-granted, the objective and the natural, the landscape masks the artifice and ideological nature of its form and content. Its history as a social construction is unexamined' (Duncan, 1990, p.19).

By contrast, Duncan and Duncan (1988, p.119) argue that the post-structural approach to text, and by implication landscape, is in some ways inadequate and 'nonsocial in that it posits an anonymous intertextual realm of interacting texts divorced from the historical, social and political processes by which interpretations of text are negotiated, contested and maintained'. There is here a sense that acceptance of an 'infinite multiplicity' of textual meanings would be antithetical to the project of critical inter-pretation. Thus 'although it is important to recognise the instability of meaning, it is equally important to realise that this plurality is finite' (ibid.). Explicitly then, Duncan and Duncan (1992, p.29) feel compelled to state that, 'we would argue that Barthes' critiques from his *Mythologies* period are particularly compelling for the geographer and other social scientists'. For within this rubric, 'virtually any landscape can be analysed as a text in which social relations are inscribed' (Duncan and Duncan, 1988, p.123). And thereby, 'the ideological aspects of landscapes as texts can be *unmasked*' (ibid., p.125, my emphasis).

---

**Box 3.2  JAMES AND NANCY DUNCAN – 'CAN'T LIVE WITH THEM, CAN'T LANDSCAPE WITHOUT THEM'**

'Landscapes are produced and maintained in ways that are largely unseen by those who happen to drive past. . . . Deeply embedded in the landscape are human costs invisible to the eye' (Duncan and Duncan, 2003, p.89).

continued

Many influential analyses of landscape as a way of seeing, or as a text, have had a historical flavour, and have focused upon, for example, contexts such as Renaissance Italy (Cosgrove, 1985) or nineteenth-century Britain and America (Daniels, 1993). Some of James Duncan's own best-known work is on the subject of landscape in colonial Ceylon (Sri Lanka) (Duncan, 1990). Recent work by Duncan and Duncan (2003, 2004), however, focuses upon contemporary American suburbia, and supplies a strong account of the cultural politics of landscape.

In 'Can't live with them, can't landscape without them', Duncan and Duncan's (2003) topic is the neighbouring suburban towns of Bedford and Mount Kisco, situated in the commuter belt to the north of New York City. The houses and gardens of the more affluent of the two, Bedford, are increasingly serviced and maintained by Latino immigrants, who themselves live in Mount Kisco. 'Bedford Village', Duncan and Duncan (2003, p.89) write,

> is considered by its residents to be an idyllically beautiful landscape of gently rolling hills. Tall maples and oaks overhang dirt roads lined with stone walls and wild flowers. Although they are hidden from view, the hilltops are dotted with late nineteenth and early twentieth century mansions, obscured by tall trees and approached by long winding gravel driveways. The aesthetic value of having a rural landscape is seen by most all of the residents of the town as unquestionable.

This idyll, however, has a rather darker side. First there is the issue of straightforward economic inequality: 'Latino immigrants have been drawn to places like Westchester County where they fill a growing niche in the local service economy, especially in landscaping. [But] . . . the jobs are typically seasonal, non-union, and low pay, with few benefits and little security' (p.91). And, second, on a different level we can think about how this economic situation intersects with questions of lifestyle and aesthetics:

Bedford is a site of aesthetic consumption practices in which the residents derive pleasure and achieve social status by preserving and enhancing the beauty of their town. They accomplish this through the use of exclusionary zoning, stringent environmental protection legislation, and the exploited labor of recently arrived Latino dayworkers. A class aesthetic based in such ideologies as localism, antiurbanism, antimodernism, anglophilia, and romanticism also underlies and lends a political dimension to the desire to live in a beautiful place such as Bedford.

(p.9)

The point which Duncan and Duncan wish to stress, therefore, is the close intermeshing of aesthetic, economic and political questions in the Bedford/Mount Kisco landscape. As they note, with regard to Bedford's pastoral beauty,

this might not have any significant social consequences if Bedford *were*, in fact unique. However, a great many of New York City's northern suburbs are characterized by similar aesthetic and exclusionary practices. Cumulatively, these practices become in effect subsidies to the rich that have the effect of reducing available land for the potential development of affordable housing and contributing to the dearth of rental housing in northern Westchester County and thus to the exorbitant rents the labourers are forced to pay.

(p.91)

A key issue here is the distinction between public and private spaces, and the corresponding patterns of behaviour that differing cultures associate with each. As Duncan and Duncan note,

the perception of an invasion [in Bedford] can be explained in large part by conflicting cultural conventions of public space based in an ethnocentric and class-based aesthetics. This has created a paradoxical situation in which those whose

continued

labour maintains Bedford's landscape aesthetic are themselves considered an unaesthetic element of the streetscape of
neighboring Mount Kisco where Bedford residents habitually
go for shopping and services.

(p.91)

This point can also be made with reference to the residential
landscape of Bedford's suburbs:

A suburban community for many Anglo-Americans is an exclusive, semi-private space where people of like minds, incomes,
and similar tastes do not so much interact as maintain
similarly aestheticized private lifestyles. The presence of
racially marked outsiders offends the aesthetic of homogeneity
necessary to the maintenance of such a community. It is not so
much the actuality of the presence of Latinos in the area as
their *visibility* that disrupts the spatial/moral order of suburban
society. To put it bluntly (as many of our informants did), the
presence of the Latino day labourers on village streets is
thought to 'spoil the *look* of the landscape'.

(p.94)

And underlying these spatial conflicts, as Duncan and Duncan
wish to stress, there is a pastoral landscape aesthetic, transplanted
from eighteenth- and nineteenth-century European landscape
art to contemporary America, where it functions as a gauze
for wealth, privilege and private landownership. In and around
Bedford,

landscapes are treated as aesthetic productions, controlled
so that as far as the eye can see, even if one drives or rides
on horseback for many miles, one views nothing industrial
or *distasteful*. Residents of Bedford maintain the illusion of
disconnection through the spatial separation of home and
work and an aestheticized attitude that conflates images of

the English country gentleman, owner of all he surveys, with the sentimental pastoralism of the Jeffersonian American small farmer.

(p.90)

The richest people, having both the greatest resources and feelings of entitlement, attempt to control long-distance views. They often go to great lengths to ensure that nothing they see from their own property and nothing they pass by when they drive around their towns is unattractive. The pleasure they take in their property, as well as its economic value, depends greatly upon control over the aesthetic and spatial practices of a whole community. As residents of Bedford and similar towns believe, ownership of land gives them the right and responsibility to produce a town's landscapes as a coherent whole, a visual production.

(p.90)

And yet, despite feeling this right to visual control, Bedford's affluent white residents are simultaneously blind to the workings of economic and cultural forces in the landscape. They bracket off questions of aesthetics and 'look', and so,

in general they see the landscape as innocent and pride themselves on their environmental consciousness. However, the immigrants who stand on the streets of Mount Kisco are there in large part because of the increasing demand for their labor on Bedford's estates. They help to sustain Bedford's pastoral landscape by recreating and maintaining miles of centuries-old dry stone walls, planting and tending gardens, mowing lawns, and repairing and repainting country. But they don't fit the Anglo-pastoral narrative structure being created in the landscape. They don't quite look like Anglo-American farm workers; their very presence is seen as a manifestation of the suburbanization of urban racial and immigration problems.

continued

> They are, in other words, seen by Bedford residents as a very mixed blessing indeed.
>
> (p.94)
>
> In conclusion, therefore, Duncan and Duncan reiterate a theme of much of their work – that landscape must be understood as a structured assembly of aesthetics, inequality and injustice: 'Unfortunately, when one takes a sufficiently critical look at the social relations underlying many places such as Bedford and Mount Kisco, it becomes clear that there is often an equally unhappy history and continuing social injustice deeply structured into the beauty of the landscape' (p.98).

### 3.5.3 Writing/worlds

In what it actually wants to say about the status and function of landscape, therefore, the 'textual' approach finds itself in close agreement with the conclusions of the 'landscape as veil' perspective. At the same time, however, there does occur a distinct shift of emphasis as regards the epistemological status of landscape representations. This shift is rehearsed in Duncan and Duncan's (ibid., p.120) argument that, 'as geographers the textualised behaviour that concerns us is the production of landscapes; how they are constructed . . . and how they act as a mediating influence, shaping behaviour in the image of the text'.

In this quote an interpretative principle is being articulated. In contradistinction to realist understandings of the relationship between words and things, signs and their referents, this principle states that cultural meanings, in the form of texts, landscapes, images, are constructed within the discursive realm. In this 'constructivist' sense, all meaning is intertextual, including that produced by critics. And the same manoeuvre necessarily implies that all meaning is always already representational. Epistemology (knowledge of the world) and ontology (the world itself) are thus conflated together, the 'world itself' being constituted through images of the world. And so, in this understanding, discourses and representations (for instance, landscape texts or images) 'are what the world is made of, really' (Matless, 1992, p.41).

Again at this point distinctions between studies of landscape per se and 'new' cultural geography more generally become blurred. During the 1990s, as 'textual' metaphors became widespread, it is possible to argue that constructivism was, briefly, close to being a paradigm within cultural geography, especially following the publication of two influential anthologies: Barnes and Duncan's (1992) *Writing Worlds*, and Duncan and Ley's (1993) *Place/Culture/Representation*. The editors' introduction to the latter collection sought, for example, to situate the programme of cultural geography squarely within the 'crisis of representation' proclaimed across the social sciences at that time (see, for example, Clifford and Marcus, 1986). A strongly constructivist position is also adopted in the introduction to *Writing Worlds* (Barnes and Duncan, 1992), a collection, notably, explicitly framed in terms of landscape. Here, the editors assert that, 'there is no preinterpreted reality writing reflects. . . . There is only intertextuality . . . [and] . . . the consequence is that writing is constitutive' (pp. 2–3). And thus following a line quite close to that of the later Barthes, the editors argue that, '"Text" is also an appropriate trope to use in analysing landscapes because it conveys the inherent instability of meaning, fragmentation or absence of integrity, lack of authorial control, polyvocality and irresolvable social contradictions that often characterise them' (ibid., p. 7). This could be read as an attempt to synthesise elements of a 'constructivist' and an 'ideological' approach to landscape. Yet 'just as written texts are not simply mirrors of a reality outside themselves, so cultural productions such as landscape are not "about" something more real than themselves . . . although not referential, such practices of signification are intertextual' (ibid., p. 5). The signifying work done by landscape texts and images, in other words, now relates only to other landscapes, and it is within this intertextual relation that landscape carries out its ideological, exclusionary function. In one sense, therefore, the metaphor of 'landscape as text' further reinforces the cultural Marxist emphasis upon landscape *as* representation. The 'physical materiality' of landscape, the lodestone of a Sauerian cultural geography, is further problematised.

Yet at the same time it is possible to argue that, in attempting to combine intertexual and ideological readings of landscape, Duncan and Duncan (1988) seek to use *landscape itself* as a *material* ground. They write, for example, that 'landscapes may be seen as transformations of social and political ideologies into *physical form*' (p. 125, my emphasis). Elsewhere, Barnes and Duncan (1992, p. 5) speak of landscape's 'fixity', and Duncan (1990) refers to landscape as 'an ordered assemblage of objects' (p.17), and as 'an objec-tifier par excellence' (ibid., p.19). Most plainly, he admits to 'an impatience with groundless idealism. Ideas take place on earth' (ibid., p.15), and here it is perhaps worth noting that the intertextual approach is more directly related, in terms of lineage, to 'old' cultural geography than the cultural Marxist tradition. In attempting to chart a terrain between intertextual principle and ideological interpretation, landscape itself is heralded as pro-viding the materiality, the historicity, even the sociality, which are regarded as being somehow absent from the former. It forms the phenomenal, empirical ground upon which textual orderings are made manifest.

## 3.6  FEMINISM AND PSYCHOANALYSIS: LANDSCAPE AS GAZE

### 3.6.1  Landscape, gender and feminist geographies

Throughout this chapter I have sought to link up specific studies of landscape with the wider emergence of new cultural geographies, espe-cially in the UK, through the late 1980s and early 1990s. The Iconography of Landscape (Cosgrove and Daniels, 1988) and Writing Worlds (Barnes and Duncan, 1992), for instance, would be key citations in any history of cultural geography. Other sources of inspiration for the cultural turn had relatively little to do with landscape, however – Jackson's (1989) well-known Maps of Meaning, notable for bringing the insights of British cultural studies regarding class, race and gender into geography, would be a case in point. And other critical and intellectual movements might be understood as being central to the cultural turn while still working on a much broader stage. An example of this would be the introduction into human geo-graphies of varied forms of feminist thought and critique, and in this section I want to consider how some feminist critical analyses have worked, in particular, through issues of landscape, gender and visual rep-resentation.

One key concern for several feminist analyses has been the symbolic and material gendering of urban landscapes within Western societies that have a patriarchical, male-dominated basis, and this leads into the concomitant, broader question of how such gendering interweaves with the histories and politics of the built environment more generally. Processes of urbanisation and industrialisation produce a differentiated urban landscape, it has been argued, in which public and work spaces have traditionally been coded as masculine, and private and domestic spaces as feminine. Thus the landscape of towns and cities reflects a gendered division of labour under capitalism and patriarchy, and, more sharply, renders particular places and times as fearful and threatening for different groups of women (Valentine, 1989). Monk (1992), Bondi (1992, 1998) and Domosh (1996) also attend to the complex gendering of the urban landscape, for example through its symbolic coding as masculine via statues and monuments dedicated to male public figures and military leaders, through patterns of retailing and consumption and through processes of gentrification and restructuring. Such analyses link up issues of landscape and gender with various other concerns, for instance, concerns around geographies of work, memory and consumption. However, in terms of the specific focus of this chapter upon landscape conceived as an ideological cultural form, a particular way of seeing the world, the work of the feminist cultural geographer Gillian Rose (1993) is most pointed.

## 3.6.2 Landscape as masculine gaze

Rose (1993) presents an astringent critique of geographies of landscape, both 'old' and 'new' cultural. She argues that while they offer themselves as critical analyses of the visual, many remain nonetheless wilfully blind in relation to their own forms of visuality. Especially those which adopt a textual metaphor. Barnes and Duncan (1992, p.9), for example, write that 'landscape possesses a similar objective fixity to that of a written text'. In the previous section it was argued that their desire to cast landscape in such terms arose in part from an anxiety about the perceived lack of empirical and material foundations within an intertextual or constructivist reading of landscape. But, in highlighting this phrase, 'objective fixity', with a different critical purpose, Rose argues that 'the notion of solidity is necessary in order to imply the possibility of certain knowledge about landscape' (1993, p.100). It allows critics to speak of landscapes as contested terrains and yet present their own interpretations as uncontaminated by contestation. Cultural geographers therefore avoid confrontation with the issue of their own *visual pleasure* in landscape. Thus 'the metaphor of landscape as text works to establish an authoritative reading, and to maintain that authority whenever emotion threatens to erupt and mark the author as a feeling subject' (*ibid.*, p.101).

This invocation of a 'feeling subject' here highlights Rose's central thesis. The landscape way of seeing is, she argues, a particularly *masculine* visual gaze. In this context, 'textualising landscape is an attempt to deny the phallocentrism of the geographic gaze, while also establishing a specific masculinity as the norm through which to access visual knowledge' (*ibid.*).

For Rose, therefore, the specific metaphor of 'landscape as text' is part of a wider and ongoing process, wherein certain masculinities and forms of vision/gazing constitute the basis of what is counted as true and proper geographical knowledge. Two related ideas further work to perpetuate this configuration. First, 'historically in geographical discourse landscapes are often seen in terms of the female body and the beauty of Nature' (*ibid.*, p.87), a relation cognate in one sense with a Western mythological tradition of identifying the figure of 'earth' with that of 'mother' through tropes of cyclicality and fecundity (see Plumwood, 1993). In another sense this relation is heavily sexualised; the female body-as-landscape forms a metaphorical terrain of desire. This metaphor is evident, for example, in processes of European exploration and travel through non-European space, where, time and again, 'women represent the enticing and inviting land to be explored, mapped, penetrated and known' (Rose, 1993, p.92).

Two dualities are being woven together here: female/male and nature/culture. The weaving works such that the 'female' becomes associated with the 'natural', the 'male' the 'cultural'. And a second discursive formation further secures and purifies these distinctions – the distinction between *mind* and *body*, which as Rose (1993, p.88) notes, further involves a separation of 'the seeing intellect from the seeing eye'. This separation, fundamental to the Western scientific tradition of detached observation, frees the masculine gaze of geographers upon the landscape from any sensual, fleshy association. Their vision of landscape is thus able to become (or be presented as) the disembodied gaze of 'objectivity'.

Yet at the same time a central tension persists within this gaze, a tension between 'rational' *visual knowledge* and 'the recurring but uneasy *pleasure* that geography finds in landscape, acknowledged but never addressed' (Rose, 1993, p.101). For Rose, however, to simply indicate the elided persistence of the sensual and pleasurable within a discourse of objectivity is insufficient. Instead the problem is to outline the processes through which the 'pleasure' of 'looking at landscape' is itself associable with specifically masculine subjectivities.

## Box 3.3  CATHERINE NASH – 'RECLAIMING VISION'

'Feminist critiques of the concept of landscape within cultural geography have made its use questionable' (Nash, 1996, p.149).

In 'Reclaiming vision: looking at landscape and the body', Catherine Nash engages with issues of landscape, gender and visuality, in a reading which at once amplifies and adds further complexity to Gillian Rose's (1993) influential arguments regarding landscape as a specifically masculine way of seeing. Nash's aim is to 'reconcile a feminist approach which retains the idea of landscape as a focus of substantive and theoretical concerns, despite feminist critiques of the masculinity of the landscape tradition within geography'.

'Rather than simply assert the oppressive nature of images of feminised landscapes or of women's bodies as terrain', Nash continues,

> it is necessary to engage with them to disrupt their authority and exclusive pleasures and open up possibilities for difference, subversion, resistance and reappropriation of visual traditions and visual pleasure. . . . Pleasure in research, writing or looking at landscape and the body is political, but this does not render this representation or vision automatically unacceptable.
>
> (p.149)

With this position in mind, Nash aims to open up cultural geographies of vision, knowledge, power and pleasure by considering the specific case of representations, by female artists, of the male body *as* landscape. Thus, she asks: what does it mean to depict the male body as landscape?

> Does it simply replicate masculine vision? Or does it free looking from its automatic implication in patriarchal and phallocentric representations and powers? Can it suggest not

continued

only other forms of feminine sexuality but other versions of masculine sexuality as passive and desired rather than active and desiring? Can it point to ways of acknowledging the erotics, pleasure and power of landscape imagery while relating to landscape in less oppressive ways?

(p.153)

Nash's article seeks to give a positive answer to this last question through an examination of two works by two female artists – a photograph, *Abroad*, by Diane Bayliss, and a video installation, *Inis t'Oirr/Aran Dance*, by Pauline Cummins. The first of these is discussed as especially germane to notions of landscape as a gendered way of seeing.

*Abroad* depicts a naked male torso, genital and thighs as though these were a topography of rolling hills and fields being viewed from afar. The edge of the naked body forms a horizon, above which a sky filled with clouds fills the frame, thereby furthering the 'landscape' illusion. Nash suggests that 'while the title may imply a representational practice outside existing representational systems, importantly *Abroad* is neither fixed within nor escapes from traditions of depicting the body or landscape, but is powerful because it plays on existing traditions' (*ibid.*, p.159).

The body in the photograph is aestheticised and sexualised through the landscape format. The spatial organisation of the image creates a vanishing point and visual focus centred on the genitals. The body is truncated [headless] in order for the allusion to landscape to work, and thus there is no returning gaze . . . significantly, the faceless body may allow the representation of female sexual and aesthetic enjoyment of an anonymous male body rather than love of a known man within conventional constructions of femininity. It is both an intimate image and distanced since the viewing position implied is both one of closeness to the body and distance from the landscape.

(pp.159–160)

Continuing this analysis, Nash notes that,

> in *Abroad* the penis is flaccid and as such it does not reproduce the tropes of 'normal' masculinity. In this landscape the penis does not break the skyline of the body's gentle contours. If male power is dependent upon the construction of masculinity which is in turn a product of an understanding of the male body as inviolable, impenetrable and whose boundaries are hard-edged and distinct, these representations of the penis as flaccid may disrupt the stability of unequal social relations based upon normative and essentialist understandings of femininity and masculinity.
>
> (p.161)

In sum, *Abroad* is an image which cuts across and though normative cultural dualities in which masculinity is coded as active and viewing, while femininity is coded as passive and viewed. Thus, Nash argues, such images indicate the capacity for the landscape way of seeing to be produced and understood otherwise. This runs slightly against the position advanced by the feminist geographer Susan Ford (1991) in which masculine distance and objectivity is to be redressed by 'acknowledging emotion and celebrating landscapes of intimacy' (*ibid.*, p.156). The difficulty, as Nash remarks, is that such a position, presented starkly, may tend to in fact reproduce stereotypical images of women as somehow essentially emotional, embodied, nurturing and caring. Her take on landscape moves in a different direction:

> In reclaiming pleasure in vision and landscape for a feminist cultural geography, rather than turning to ideas of a feminine and body-centred identification with nature, it may be productive to think through a range of potential identifications with landscape or nature, including those of indifference or disinterest. . . . Rather than argue that the politics of representation or visual pleasure can be assessed by reference to a

continued

> male or female gaze, it is more useful to think of a multiplicity
> of shifting viewing positions, gazes or ways of seeing.
>
> (pp.156–157, p.159)
>
> What these insights suggest, coupled with the analysis of *Abroad*, is
> that 'looking is not always necessarily powerful and oppressive, nor
> is distanced vision always an appropriation. . . . To picture the
> male body as landscape is not simply to invert customary gendered
> positions within a scopic regime, but to suggest other pleasures
> and subject positions which are not determined by this regime'
> (p.161).

### 3.6.3  Voyeurism and narcissism

Within feminist studies of the visual arts in particular, these processes have
been most commonly approached from a psychoanalytic perspective. In a
noted text in this area, the art historian Griselda Pollock (1988, p.147),
situates this manoeuvre as both a complement and a challenge to the
cultural Marxist and iconographic histories of art discussed earlier in
the chapter (Section 3.3):

> The use of psychoanalytic theory not only provides some inter-
> pretative tools for understanding the obsessive preoccupation with
> images of women . . . but also shifts attention away from icono-
> graphic readings to the study of the process of the image, what is
> being done with it and what it is doing for its users.

As Laura Mulvey (1989, p.14) demonstrates in a renowned polemical
critique of 'the way the unconscious of patriarchal society has structured
film form', this processual relation between the image and those who gaze
upon it is psychosexually charged. Drawing upon both Freud and Jacques
Lacan, she outlines two psychic structures enacted by the gaze. The first of
these is scopophilia, or voyeurism, the pleasure derived from a gaze upon
an 'other' which is not reciprocated. The object of this gaze is thus literally
'objectified', and in this way voyeurism constitutes 'the erotic basis for
pleasure in looking at another person as object . . . taking other people as

objects, subjecting them to a curious and controlling gaze' (Mulvey, 1989, pp. 16–17). In a patriarchal, that is, male-dominated society, Mulvey argues, voyeurism occurs as a masculine gaze, producing objectified images of the female ('woman as image, man as bearer of the look' (ibid., p.19). This is a gendered opposition which further codes the male as *active*, and the female as *passive*.

In addition to being voyeuristic, however, in psychoanalytic terms the visual gaze is also fundamentally *narcissistic*. In a celebrated study, the French psychoanalyst Jacques Lacan (1977) argued that narcissism, or reflection upon one's own image, is a constitutive act in the production of a sense of self – what he terms 'the mirror stage of ego formation' – the moment when the infant recognises its boundedness through recognising its own reflection in a mirror. From a feminist psychoanalytical perspective, the crucial aspect of this formative 'look' of masculine identity and subjectivity is that it designates the 'female' in terms of *lack* and *absence*; what Freud famously called the 'castration complex'. Thus, for the gaze, the female 'connotes something that the look continually circles around but disavows: her lack of a penis, implying a threat of castration and hence unpleasure' (Mulvey, 1989, p.21). As Griselda Pollock (1988, p.139) argues, this coding of the 'female' as 'lack' produces two conflicting visions: 'one is the compensatory fantasy of the pre-Oedipal mother, still all-powerful, phallic; the other is the fantasy of woman not only as damaged, but as damage itself'. The response to these contradictory images is either fetishism, where the image is transformed into an objective spectacle of desire, or what Julia Kristeva terms *abjection*, where the 'female' is coded as being anterior to the formation of identity, as being, in fact, 'the means whereby the subject is first impelled towards the possibility of constituting itself as such – in an act of revulsion, of expulsion' to the margins, to the very vanishing points of the space of representation (Burgin, 1996, p.52).

Gillian Rose (1993) sees in such ideas a correlation with, and explanation of, concepts of landscape as a 'way of seeing', or *gaze*. Mulvey's (1989, p.20) account of narcissistic identification in the cinema states that,

> The active male figure (the ego ideal of the identification process) demands a three-dimensional space corresponding to that of mirror recognition. . . . He is a figure in a landscape. Here the function of film is to reproduce as accurately as possible the so-called natural conditions of human perception.

And drawing upon the same analytic, Rose (1993, p.108) argues that

> If – contingently – heroes in landscapes correspond to the coherent active subjects that we (mis)recognise in the mirror, this process surely accounts for some of the satisfaction of fieldwork for geographers. They see themselves as the ego-ideal hero in a landscape.

More specifically, in terms of visual knowledge and representation, she argues that 'geographers try to repress their pleasure in landscape by stabilising their interpretations as real; but that knowledge is, in its need for critical distance, implicated in the pleasures of voyeurism. They try to win knowledge through intimacy with the land, and their intimacy becomes narcissistic' (ibid.).

The masculine gaze upon landscape is therefore in its voyeuristic mode 'a gaze torn between pleasure and its repression' (ibid., p.103) – yet such repression, in the form of an appeal to 'distance', 'objectivity' and 'neutrality', nonetheless presupposes the very voyeurism it attempts to deny. At the same time, the narcissistic mode of identification with the image of the 'heroic fieldworker' also subsists within the aesthetic coding of 'woman' as 'nature', such that 'pleasure in landscape comes partly from its seductively sexual vision of narcissistic reunion with the phallic mother' (ibid., p.105).

In conclusion, as Nash (1996, p.155) summarises,

> for Gillian Rose, studies within the landscape tradition in British cultural geography of the class relations constructed and reproduced within nineteenth-century English rural landscapes, and virtual silence on issues of gender, signifies not only the omission of gender issues and critical engagement with patriarchal power, but a deep ambivalence regarding the object of study. The ambivalent pleasures of landscape within cultural geography draw not only on a complex discursive transcoding between woman and nature, but also on a specific masculine way of seeing.

As Rose herself (ibid., p.108) states:

> The intersections of voyeurism and narcissism . . . structure geography's gaze at landscape. The gaze which identifies lack in the compelling vision of Nature as woman maintains a voyeuristic

distance from that which represents lack; but it is also compelled to gaze and gaze again through its desire to interpellate itself through the feminine.

## 3.7 DISCUSSION AND SUMMARY

Three approaches to landscape have been outlined in this chapter. Each of these advances upon a definition, of sorts, through providing a critique of its primary theoretical legacy: 'humanistic' geography in the case of the 'landscape as veil' school; Sauerian cultural geography for those who view 'landscape as text'. Equally, Rose's (1993) analysis of the landscape as 'gaze' is able to reflect critically upon both 'text' and 'veil' as interpretative metaphors, through aligning them with long-standing visual traditions in geography. And each of these three approaches also advances its own distinctive set of interpretative supports and procedures: in turn, the cultural Marxisms of Berger and Williams; the structuralist textualities of Barthes; the psychoanalytics of Freud and Lacan. Each approach is therefore different, distinctive in its legacies, techniques and concerns. But I want to suggest here, by way of summarising and concluding this chapter, that certain crucial similarities may also be witnessed across all three – regularities of both method and definition with respect to landscape.

First, most obviously, landscape is defined by all three as quintessentially *visual*. This, it might be argued, is unremarkable, self-evident, given that within its primary domains of explication (the histories of art and geographical practice), landscape has forever most commonly been associated with a particular 'area of space' as *seen*. Landscape, therefore: a unit of visual space. This definition is variously re-inscribed within the cultural geographies discussed here: landscape as a painterly patina, as a text to be read, as a form of visual desire. The crucial distinction which all share, however, is best described as an interest not in the 'seen', but in 'ways of seeing'. Landscape in 'new' cultural geography came to refer to the outcome of particular *visual processes* of description and symbolisation. As an 'outcome' (a painting, a garden, a poem), landscape perhaps retains a certain solidity – it may still be an object or an environment – but this is a point that can be quickly passed over, for what is crucial, what is essential, are the specificities of the 'way of seeing' which constitute any landscape's meaning.

Landscape is a visual image of cultural meanings. It is therefore both the product and the token of particular cultures, particular knowledges and subjectivities. Cosgrove, Daniels and Duncan emphasise that landscape is a 'way of seeing' cognate with the visual gaze of European socio-economic elites. Cosgrove and Rose highlight visual landscape's association with the detached gaze of both empirical and rational science. Rose demonstrates that, as an object of enquiry, landscape is constructed by a masculine, voyeuristic and narcissistic gaze. Daniels (1993) illustrates landscape's symbolic function in the sustenance of particular visions of national identity. And, as will be discussed further in the next chapter, landscape may also be spoken of as an 'imperial' gaze, as the visual modus operandi of European explorers, artists and cartographers, as 'this haughty gaze that has surveyed and appropriated the world' (Duncan, 1993, p.40). Landscape therefore becomes for new cultural geographers the visual medium through which they can provide critical interpretations of social and cultural formations.

This shift from 'seen' to 'seeing' also serves to index a second alignment between the three approaches to landscape outlined here: all of them see landscape as *representation*. Again this is perhaps unremarkable; again both geographers and art historians have always sought to describe what a given landscape stands for, in other words what cultures and histories it expresses or symbolises. The distinctive gradient of 'new' cultural geographies, however, lay in the ascending status they accorded to representation and representational practices. This gradient may be witnessed substantively, methodologically and theoretically. On a substantive level, 'landscape' in 'new' cultural geography referred no longer to a physical environment, but rather to images, paintings, idioms of description in travel accounts. This shift had a methodological correlate, in that the concern of geographers was no longer *description*, but rather *interpretation*. Instead of producing representations of landscape, 'new' cultural geographies of landscape took these as their primary object of concern. And, finally, this interpretative turn in terms of method, both heralded and reflected the theoretical stance that became dominant within cultural geographies more broadly through the 1990s, a stance perhaps best described as 'critical constructivism'. This stance held that meaning and knowledge – the discursive structures of socio-cultural systems which the critic interprets – belong to, and emanate from, the symbolic domain of images, signs, texts, representations.

As Ulf Strohmayer (1998, pp.105–106) noted with reference to these sorts of ideas, 'the constructedness of representations could well be the bond unifying most of the work currently undertaken within geography . . .

constructedness thus forms a kind of "pragmatic" response to the "crisis of representation"'. In a rather paradoxical sense, therefore, the critique of mimesis and correspondence across the social sciences resulted in representations and their construction assuming an even higher status as the primary object of enquiry. The central point here is that whilst, for new cultural geography, all knowledge and meaning and representation was constructed, and therefore contingent, some forms of representation were particularly cogent and power-laden and thus deserving of critical attention. And one especially noteworthy regime of representation was the landscape way of seeing, a gaze projected out onto the land, a vision of authority and ownership, the mind's eye of certain knowledge systems, vested interests and desires. The task of this chapter has been to outline in detail the critical arguments and theorisations which underpin this definition of landscape as a way of seeing. In the following chapter I will discuss how this idea was taken up and developed anew by cultural geographers through the 1990s.

# 4

## CULTURES OF LANDSCAPE

### 4.1 INTRODUCTION

Following on directly from the discussions of the previous chapter, this fourth chapter considers the diverse – and diversifying – range of landscape literatures which emerged in anglophone cultural geography through the course of the 1990s. In some ways, therefore, this is an evolutionary story: a story about how a central critical understanding of landscape as a way of seeing – landscape as a visual representation of cultural meaning and power – was extended, critiqued and developed by a generation of geographers emerging in the years following the publication of key texts on landscape by writers such as Cosgrove, Daniels, Duncan and Rose. In particular, the idea that landscape is a 'gaze' imbued with Western cultural attitudes and perspectives, a gaze at once contingent yet powerful, is one which has proved durable, productive and lastingly influential. The notion of 'ways of seeing', with its critical imbrication of landscape, visuality, power and ideology, remains cogent today. Thus, in part, this chapter will detail how this idea has been used and applied in a series of different geographical and historical contexts. To reiterate points previously stressed, alongside a conception of culture as a signifying system drawn from semiotics, and the widespread adoption of an anti-essentialist and anti-foundationalist episte-mology, the idea that landscape is best conceived as part of a 'constructed' and circulating system of cultural meaning, encoded in images, texts and discourses, becomes a central theme in what might retrospectively be called the critical-constructivist paradigm of 'new' cultural, social and historical geographies.

But it would be quite wrong to suggest that what we see in cultural geographies of landscape through the 1990s is simply a continuation and further elaboration of readings of landscape as veil, text and gaze. In other ways, as this chapter will show, this period witnessed a series of reactions against, amendments to, moves away from and significant re-articulations of the definition of landscape as a way of seeing the world.

One way of understanding these shifts is to place them within the broader context of evolving intellectual currents across the arts and humanities. The first generation of 'new' cultural geographers drew inspiration from and situated themselves within the traditions of cultural Marxist ideological critique, as exemplified by John Berger and Raymond Williams, and more broadly structuralist continental analyses, whether Barthean semiotics or Lacanian psychoanalysis. One element common to all of these is a tendency to stress the façade-like quality of cultural manifestations, such as landscape, and thereby to focus upon 'unmasking' the more-or-less systematic operation of structures of power and authority behind such façades. By contrast, the second generation, so to speak, of new cultural geography, is characterised by the ascendancy of *poststructural* perspectives. In particular, the poststructuralist writings of Michel Foucault (1977, 1989, 1991b) on discourse, power and practice, and of Jacques Derrida (1973, 1976, 1978) on language and epistemology, had a decisive influence upon cultural geography as a whole through the 1990s. In the specific case of landscape, this influence is most clearly evident in the emergence first of a raft of studies of what David Matless (1998) – a key figure in these shifts – terms *cultures of landscape*: studies in which the focus is not so much upon specific places, texts or works of art, as upon more multifaceted cultural movements, debates and practices in which landscape circulates both materially and symbolically, for example debates over citizenship, identity, health, planning and ethical conduct generally. Second, the influence of poststructuralist thought was paramount in the formulation of both post-colonial studies and visual culture studies in the 1990s. Enrolled in these agendas, cultural and historical geographies up to the time of writing have explored cultures of landscape – for example, styles of moving and looking, forms of land settlement and management, aesthetic ideals, exploratory discourse, scientific visual practices – as key elements within Western visual cultures, and within histories of travel, exploration, colonialism and imperialism.

This chapter will trace these developments in landscape studies through two substantial sections (4.4 and 4.5) – sections dealing first with what I will discuss as a Foucauldian understanding of landscape, subjectivity and practice, and second with intertwined interrogations of landscape, visuality and colonial and imperial discourse. In the course of these sections, I shall also spend some time elaborating the principles and epistemological positions underlying such poststructural analyses, in particular notions of discourse, power and subjectivity as elaborated in the work of Michel Foucault. Prior to these discussions, however, the chapter shall begin by looking at what has been a perhaps more parochial geographical debate. This concerns the perennial question of the *materiality* of landscape, and in particular the status of materiality, action and production within a paradigm emphasising the symbolic meaning and function of landscape. Following a discussion of the various material anxieties infecting cultural geographies of landscape in the mid-1990s (Section 4.2), the chapter will consider the overtly materialist and productivist responses offered by some, mainly North American, landscape geographers (Section 4.3).

## 4.2 MATERIAL ANXIETIES

Landscape, in the exemplary form of works of landscape art and the landscaped gardens of country estates, was understood by new cultural geography to be a *representation* or *symbolisation* of particular subjectivities, of particular cultural attitudes and values. For example, as the previous chapter discussed, landscape could be scripted as a masculine gaze, or as the way of seeing of a landowning elite, or as an aestheticised picture of the natural world and culture–nature relations. In each of these understandings, critical interpretation works through a referential movement from the landscape 'itself', as it were, to a more ideational and symbolic level of cultural meaning and signification. What is really important in this paradigm, therefore, are the ideas represented by landscapes. But, at the same time, throughout the 1990s currents of unease circulated within cultural geographies of landscape, currents produced by a persistent feeling that such a focus upon representational, symbolic and iconic meaning was problematic in so far as it tended to elide the *materiality* of landscape itself and also the cultural, political, economic and environmental relations enacted through and within landscapes. This unease was previously highlighted as an element in the 'landscape as text' school, evident in James Duncan's 'impatience with groundless idealism. Ideas take place on earth' (1990, p.15). It also implicitly informs a review of the literature by the same author (Duncan, 1995), which claims a contrast between 'painterly' and 'material' approaches to landscape.

A similar distinction between landscape as land/territory and landscape as scenery is at the heart of Kenneth Olwig's (1996) detailed attempt to reclaim the landscape concept from its new cultural association with art, aesthetics and hence the symbolic values and tastes of powerful elites. For Olwig, the high tide of the 'scenic' understanding of landscape, evident in works such as Cosgrove and Daniel's (1988) *Iconography of Landscape*, is simply 'the most recent step in its disciplinary dematerialisation' (1996, p.630). He traces this process back to the American geographer Richard Hartshone's (1939) influential rejection of Carl Sauer's advocacy of cultural landscape as the central organising concept of geography, in favour of 'a science of region and space' (ibid.). In place of dematerialisation, Olwig proposes a return to what he terms the 'substantive nature of landscape', that is, to a landscape that is real as opposed to artistic and, furthermore, real in a specifically *legal* sense: 'by substantive I mean "real rather than apparent" . . . I am also concerned with landscape as a "real" phenomenon in the sense that the real relates to things in law, especially fixed, permanent or immovable things (land tenements)' (ibid., p.645). Through this, so to speak, real-estate definition, Olwig seeks to reclaim a measure of Sauerian cultural geography, by linking its intellectual heritage with a northern European history in which the term landscape is associated with locality, community and customary law (Chapter 6 will comment further on the relation between landscape and legal geographies). Here, therefore, landscape suggests a vernacular 'people's' history of cultural use, value and transformation, and not an elite way of seeing.

Concerns regarding overly symbolic, artistic and theoretical tendencies within new cultural geographies also surface at this time in the broader field of writing on environmental issues and scientific understandings and framings of the natural world. For example, David Demeritt (1994, p.172), drawing a contrast between work from cultural geography and environmental history, argued that the emphasis upon metaphorical meaning and reference in the former had the effect of silencing nature:

> Landscape metaphors of cultural production have, both in theory and practice, served to make nature ephemeral and epiphenomenal. These metaphors treat nature as a blank page or an empty stage on which the drama of culture is written and acted out. . . . In moments of metaphorical extravagance the material 'reality' of landscape disappears altogether.

Although he himself subsequently pursued much more sophisticated analyses of nature and scientific epistemologies, Demeritt's statement may be read in part in terms of a deep-rooted geographical empiricism, widely evident in characterisations of geography as a 'grounded' practical discipline, and reflecting the persistent influence of associations between landscape study, evidentiary historical scholarship and field methodologies (see Chapter 2). To put it more crudely, perhaps, there have always been elements opposed to a clear identification of human geography with the arts and humanities disciplines. In this way, as David Matless (1995a, p.396), noted in a third review of the literature coterminous with those of Duncan and Demeritt, it was most often voices opposed to the 'cultural turn' in human geography per se that sought to portray culture as 'a realm of representation into which the material world is to be converted and dematerialised'.

The fact remains, however, that issues around materiality have haunted writers committed to a cultural geographical perspective on landscape. As this chapter will go on to discuss, one response has been (re)turn to an explicitly materialist, i.e. Marxist focus on production (Section 4.3); another has been to focus on the discursive *and* material world of actions, practices and performances which work within and through landscape (Section 4.4). A third response, most evident in the mid-1990s, was to view landscape's materiality as a matter of local contexts and circumstances, physical or human, and to position such circumstances as disruptive of or resistant to wider discursive or idealist projects (for example missions to 'visually instruct' colonial populations stumbling on local resistances and reworkings (Ryan, 1994), or Antarctic surveying programmes foundering on recalcitrant climates and terrains (Dodds, 1996)). In a sense here the materiality of the landscape, understood as a complex of contextual processes and circumstances, is, for such cultural geographies, its definitive hallmark. Landscape's physical and cultural materiality becomes the very stuff of interpretative purchase, enabling detailed contextual critiques of wider discursive ensembles.

But at the same time it can be argued that such accounts in fact have a deeply reductive effect. The difficulty, perhaps, is that while materiality is clung to for reassurance and anchorage, it is simultaneously positioned by many cultural geographies as a decidedly secondary characteristic. What is primary, what is held to possess true epistemological significance, is an ideal symbolic realm, wherein 'meaning' first takes shape in the process of cultural construction which produces forms such as the 'landscape gaze'. This, then, is the actual object of critique. By contrast, the material landscape, most commonly reinscribed as 'context', is rendered purely reactive, and possesses only the inertia of substance, as a rock upon which waves of discursive meaning break.

One example is especially instructive as regards the hierarchical dualism between cultural meaning/materiality at work within many cultural geographies of landscape, coming as it does from precisely the school of thought for which materiality was a pressing issue. Duncan and Duncan (1988) use the beliefs of Australian Aborigines to illustrate the interpretative power of their understanding of landscape as text. They argue that 'aboriginal peoples have a set of oral religious texts which are transformed into the landscape' (p.122). Thus, for the Aborigines, 'a rock is a rock but also a mythic being' (ibid.). Yet this statement perhaps reveals more about the authors' beliefs than those of the Aborigines, because it performs, and presupposes, a distinction between on the one hand the rock's 'rockiness' and on the other its 'meaningful cultural significance'. In the representational or symbolic approach to landscape all material objects must have a 'but also', a significance that is anterior to, and constitutive of, their meaning as it occurs through encounter and use. For in itself the rock is regarded as an epistemological nothing, requiring the supplement of 'culture' to achieve phenomenal fullness. Meaning, in other words, must infuse the rock from the outside. Meaning must be first imagined, before material action is performed. The material landscape is in itself mute and passive, a lack without force.

## 4.3 LANDSCAPE, PRODUCTION AND LABOUR

### 4.3.1 Materialist landscapes

One possible solution to the vexed question of how adequately to account for the force of landscape as material object, context and process might be found within the very intellectual traditions that first inspired new cultural geographies. Marxism, through its various twentieth-century transformations and manifestations, has consistently been a *materialist* philosophy, in so far as its central emphasis has been upon material processes of labour and production, political praxis and struggle and commodity exchange. The 'cultural Marxism' of figures like John Berger and Raymond Williams drew attention to the ideological function of Western cultural products such as commodities, texts, artworks – and landscapes. Such products had ideological import precisely because they covered up and mystified the true – that is, unjust and exploitative – nature of social and economic relations in a capitalist system such as in the West. Here we see a more sophisticated rendition of the classical Marxist distinction between, on the one hand, a primary economic *base* of material production and exchange, and, on the other, a secondary ideological *superstructure* made up of social norms, regulatory institutions and cultural products and values. Having innovatively positioned landscape within this ideological realm, as was discussed in Chapter 3, the earliest task for new cultural geographers such as Denis Cosgrove and Stephen Daniels was to detail how landscapes, painterly and literary, functioned as glosses, façades and aesthetic veneers, designed to perpetuate existing social, economic and political hierarchies.

But is this all that can be said about landscape from a Marxist or materialist perspective? In a sequence of articles and books, cultural geographer Don Mitchell (1994, 1995, 1996, 1998a, b, 2001, 2003a), echoing and ramifying the work of other radical North American geographers and writers such as Smith (1990), Zukin (1991) and Wilson (1992), offers what is a perhaps more systematically materialist interpretation of contemporary landscape processes, particularly in a US context. This section discusses Mitchell's work in particular. Necessarily, perhaps, this work stands in an ambivalent relation to the new cultural geographies of landscape previously discussed; heavily critiquing some tendencies, but also in part chiming with them, and even further clarifying their insights. This reflects in part the fact that many of these new cultural geographies had their origin and biggest impact in British geography – the 'cultural turn' did not register in the same way in US geography, a point that will be enlarged upon in Chapter 6 via a discussion of current work on landscape by US-based cultural and political geographers (including Mitchell).

Mitchell's (1995) critique of new cultural geography is striking in so far as it echoes the very criticisms that writers such as James Duncan (1980) and Peter Jackson (1989) had earlier levelled at Carl Sauer and the Berkeley School (see Chapter 2). His argument is that, in adopting an understanding of culture as a medium or 'sphere' in which human life is lived, new cultural geography conceives of culture as an ontological entity possessing causative power, as something that really exists, in a manner markedly similar to the 'superorganic' understanding of culture proffered by 'old' cultural geography. To put this more directly, in focusing upon 'culture' per se as something that can be used to *explain* real-world events and processes, new cultural geography reifies and exalts the concept of culture in an inappropriate and unsustainable fashion. Thus, in terms of definitions of culture, 'the shift from determinant "thing" [in the Berkeley tradition], to nebulous "level" [in new cultural geography] has had the effect of further mystifying processes of social power as well as continuing to reify the essentially empty, untethered abstraction of "culture" ' (1995, p.103).

The point which Mitchell wishes to stress is that 'it is a fallacy to assume that culture has an ontological existence' (p.110). Having fallen into this trap, the approaches and concepts associated with new cultural geography perpetuate a series of mystifications:

> These ways of seeing 'culture' do not avoid reification, rather they perpetuate right into the heart of geography what are still a quite mystified set of assumptions about how social practice proceeds. And this will continue to be the case until social theorists dispense with the notion of an ontological culture and begin focusing instead upon how the very idea of culture has been developed and deployed as a means of attempting to order, control and define 'others' in the name of power or profit.
>
> (Mitchell, 1995, pp.103–104)

Mitchell's (1995) own position (and this is his paper's title) is that 'there's no such thing as culture', there is, instead, 'only a very powerful *idea* of culture' (*ibid.*, original emphasis). His argument explicitly calls for a strong return to the Marxist notion that cultural forms (art, literature, fashion, cinema – and landscape), constitute an ideological realm through which powerful economic and political interests exercise control, in effect 'duping', deceiving and rendering passive the wider population. Thus, while he is sympathetic to the original aims and procedures of new cultural geography, Mitchell argues that something has gone crucially awry: 'we have lost sight of the idea of culture as ideology. We risk abandoning the important political goals of the various "new" geographies that emerged out of the ferment of the 1960's' (ibid., p.112).

A more strongly materialist critique would reinvigorate and redirect cultural geography, establishing once again as a central focus struggles for social and economic justice, and the critical exposé of the various instruments of coercion and domination deployed by states and powerful capitalist interests. But how would this operate in practice in terms of analyses of landscape? Mitchell's answer here is to point to another relative blindspot in new cultural geography: a lack of reflection upon the question of how landscapes are produced. This is highlighted in an early paper on landscape and labour:

> theories of 'landscape as text' understand the landscape to be both an outcome and a reflection of cultural values. . . . Although these recent theories of landscape textuality are often quite sophisticated in methodology and politics, they also often suffer from a neglect of the facts of landscape production. Within this methodology, readable landscapes are already there to be decoded; the *process of their production is rarely enquired after.*
>
> (Mitchell, 1994, p.9, emphasis added)

In other words, an interpretative focus upon already 'complete' landscapes, textual, visual or material, is arguably insufficient. While such a focus might effectively reveal landscape's role as the upholder of certain ideological values, it would have little to say about the processes through which the landscape was made and produced. To put this another way, the interpretative approaches associated with writers such as Cosgrove, Daniels and Duncan only address half the story – how landscapes are consumed. For Mitchell, by contrast, the key to understanding landscapes is to consider how they are being produced.

## 4.3.2 The work of landscape

The production of landscape falls under the rubric of a Marxist/materialist understanding of how capitalism utilises the power of labour in order to transform raw materials with the aim of generating profit. For Mitchell, the process of making an actual, material, working landscape – a fruit farm, say,

## Box 4.1 STRAWBERRIES

The critical landscape geographies developed by Don Mitchell and others aim above all to highlight and examine the social and economic relations at work in the production of landscapes. This box develops a concrete example of such an analysis.

If we think of a landscape and its products – a piece of fruit such as a strawberry or tomato, say – as simply 'natural', then they can easily appear to our eyes as innocent, untouched and unproblematic entities. Objects without broader context, they seem to simply exist as parts of 'nature'. However, once we begin to think about the forces that *produce* the landscape and tomato and strawberry – things such as systems of agricultural labour and production, transnational commodity chains, buyers and sellers, transportation and storage providers, advertisers, supermarket layout designers – then it quickly becomes apparent that these are social and economic as much as natural products.

One way of thinking about this is to say that apparently natural entities or environments are, so to speak, 'socially constructed'. We project our perceptions, values and attitudes onto objects and landscapes, and so they become readable as 'maps of meaning' (Jackson, 1989). To put that another way, objects and landscapes acquire symbolic significance through being imprinted within shared and collective circuits of cultural meaning. For example in the UK strawberries might well be associated with luxury, summer, privilege, Wimbledon and so on.

Another way of approaching this issue – the way followed by Don Mitchell with respect to *landscape*, and Noel Castree (2005), among others, with respect to *nature* – is to emphasise that commodities such as strawberries or tomatoes have been produced, that is actually physically created, via sets of social and economic relations. The point to be drawn from these ways of thinking is that once we stop thinking of landscape as part of a separate, god-given nature, or as simply a neutral backdrop or setting for human activity, and begin instead to examine the ways in which landscapes

continued

are implicated within and reflective of social, political and economic circumstances, then we also begin to move from a naive and simplistic understanding of landscape towards one which is more subtle, engaged and above all *critical*.

Landscape is not the setting for human activity, it is the product and outcome of such activity. Therefore we study landscapes for what they may reveal about the nature of human social and economic relations. This argument is made forcefully by Don Mitchell (2003a, 2001, 1996) in his extensive analyses of systems of agricultural production in California. Mitchell begins one particular essay with an image of himself in a kitchen, washing some strawberries. He writes, 'the fruit I was preparing that morning – *its very shape and structure* – was the product of countless hours of labour . . . [however] . . . the strawberry, as it is being washed and sliced, says nothing about the labour that makes it' (Mitchell, 2003a, p.235, emphasis in original). In other words the strawberry is a 'social' product, just as much as a car or a computer, yet in and of itself it hides these origins and masquerades as a 'natural' object to be modified and used by humans. The strawberry is best thought of as 'dead labour': it is a commodity in which human labour is materialised and made stable.

Mitchell makes the point that this analysis also applies to, and from his perspective defines, landscape: 'by extension the landscape can be understood to be a product of human labour, of people going to work on the land to make some-thing out of it' (*ibid.*, p.238).

This might seem in some ways a self-evident observation. However, the argument that a landscape is the outcome – the frozen or solidified form – of active human labour is a Marxist and materialist one, and in drawing upon the Marxist critical tradition Mitchell's foremost concern is to expose the unjust, unequal and exploitative nature of the relationships between human beings that occur within capitalist systems such as the Californian agricultural economy. Landscapes are 'duplicitous' (Daniels, 1989), in so far as they tend to present themselves as 'scenery' or as 'nature' in a way that obscures and masks the social and economic conditions that go into their making. In this way landscapes become the secret standard-bearers of ideological values. By masking the processes by which they were created through being seen and understood in solely visual and aesthetic terms, landscapes work so as to make contingent sets of historical circumstances, and particular types of economic relations (for example relations between landowners and labourers), appear natural and normal. And as long as there is a lack of consciousness regarding this process, Mitchell argues, in other words as long as social and economic relations are not the object of debate, struggle and conflict, then landscape's primary function is to generate profit, or surplus value. This is evident in the Californian agricultural system: 'to the degree that the landscape is uncontested, to the degree that labour unrest can be stilled because of the sense that there is simply no alternative, then surplus value [profit] can be expanded. Landscape is thus a form of social regulation' (2003a, p.241).

This regulation is not merely a banal matter of rules and by-laws. For Mitchell, as a form of regulation landscape is, finally, about violence. He concludes:

The California landscape might best be described as being constituted by a series of 'points of passage' within a 'network of violence' (Mitchell, 2001) . . . real, bodily, physical violence: farmworkers' hands mangled by machinery, the gun and knife fights in the cities and the labour camps that are often part of farmworker's everyday lives; the remarkable violence visited upon migrants as they attempt to cross the border (rape, assault and murder, as well as death by exposure in the deserts and mountains); and in the violence implicit and explicit in economic dislocation, threatened starvation, and disrupted families and local ways of life in the source countries and villages of California farmworkers. Landscape – as a stage of production and reproduction – is knitted together by this network of violence. Landscape – as a 'way of seeing' that

continued

> aestheticises or erases the facts and relations of work – knits
> together this network of violence.
>
> (p.242)

or a mining town – has to be understood in terms of the axioms of classical Marxism: the labour theory of value, commodity fetishism, the need for surplus value, or profit, ongoing struggles between the interests of capital and labour. Fairly unequivocally, therefore, 'the production of landscape morphology is an essential moment in the production of surplus value [profit] in capitalism' (Mitchell, 1994, p.9). In order to fully understand the forms of land use and occupancy we see when we gaze upon any urban or agricultural scene we have to refer to systems of capitalist production and reproduction.

Of course the problem with this understanding of landscape is that it is open to the very same form of critique that Mitchell applies to new cultural geography. The underlying system of capital accumulation is here presented as just as much a totality, and is granted just as much if not more ontological existence and causal power, as any reified concept of 'culture' has ever been elsewhere. One, albeit partial, way in which Mitchell's work addresses this difficulty is through an insistence that a landscape is never finally 'produced' in the sense of being 'finished': instead landscapes are viewed as being constantly *in production*; that is open to change, alteration and contestation. This vision of a world in ongoing process further chimes with the radical dialectical politics which animates and propels materialist philosophies. Struggle and conflict become standout motifs. Thus Mitchell (1998b) endorses Zukin's (1991, p.16) arguments that landscape is 'a contentious, compromised product of society, shaped by power, coercion and collective resistance'. This notion is yet more fully voiced in Mitchell's own statement that 'social groups with differing access to power, financial and social resources, and ideological legitimacy, contend over issues of production and reproduction in place. Out of these contestations the form of the landscape is produced' (1994, p.10).

At the same time, while foregrounding social, political and economic contention as the key processual element that both produces and propels landscape forms, Mitchell's work does not aim to redeem the landscape concept, nor does he really seem to view landscapes as potentially redemptive or transformative agents. This becomes clear when we move from production to end-products – that is to *commodities*. For Mitchell, 'as "activity" landscape is always in a state of becoming; no aspect of it is entirely stabilised. Yet landscape is *also* a totality. That is to say, powerful social interests are ever trying to represent *the* landscape as a fixed, total and naturalised form' (Mitchell, 1994, p.10, original emphasis).

What Mitchell seeks to clarify, therefore, is a double reading of landscape, as something at once intimately active in the production and reproduction of capitalist social and economic relations *and* as an outcome, reflection and ideological standard-bearer of such relations. In this sense,

> 'landscape' is best seen as both a work (it is the product of human labour and thus encapsulates the dreams, desires and all the injustices of the social systems that make it), and as something that *does work* (it acts as a social agent in the further development of a place).
>
> (Mitchell, 1998b, p.94, original emphasis)

As a product – a commodity – fashioned under the rubric of modern and industrial capitalism, the work of landscape is ideological, as Berger and Williams and Cosgrove and Daniels had all each in turn suggested. For Mitchell, just as for these others writing in a materialist vein, landscape works so as to naturalise, stabilise and render apparently universal contingent social and economic relations.

> in many respects [landscape] is much like a commodity: it actively hides (or fetishises) the labour that goes into its making . . . those who study landscape representations are repeatedly struck by how effectively they erase or neutralise images of work. More particularly, landscape representations are exceptionally effective in erasing the social struggle that defines relations of work . . . the things that landscape tries to hide, in its insistent fetishisation, are the relationships that go into its making.
>
> (Mitchell, 1998b, pp.103–104)

The scope of such an analysis is potentially, perhaps imperatively, global, and in addition to his own recent elaboration of landscape in terms of 'points of passage' within much wider 'networks of violence' (see Mitchell, 2001), Mitchell's work connects up strongly, for example, with influential materialist geographies of justice and difference (Harvey, 1996), political ecology (Bryant, 2001) and uneven development and the production of nature (Smith, 1990). However, one of the interesting aspects of Mitchell's work on landscape is his tendency to focus in upon specific American material locales. Most commonly these are landscapes that have quite clearly come into being through, and then been devastated by, capitalist agricultural and industrial processes. Working landscapes: Brentwood, California; Jonestown, Pennsylvania; Butte, Montana. To conclude this section, if there is any positive recuperation of geographical notions of landscape in Mitchell's writing, then perhaps it lies in this tendency to value the local, the concrete and, explicitly, the 'ordinary'. Thus, with reference to new cultural geography's focus on landscape texts and images, he writes that 'to abandon the mechanics of landscape production at this point in a search for culture or cultural meaning is to abandon the project of understanding the ordinary just when it gets interesting' (Mitchell, 1994, p.8). He might well endorse many of the critiques of Carl Sauer and the Berkeley School approach to culture, and equally condemn the subjectivism of geographical humanism's engagement with 'ordinary' and 'vernacular' landscape (see Meinig, 1979a), but Mitchell (1998b) nevertheless stands as a latter-day, left-field representative of a North American tradition, one that insists that landscape is to be found not in rarefied or spectacular locations, but more prosaically at home and at work, in the rhythm and tare of the ordinary, even if that ordinariness precisely works to mask the processes by which it is sustained and in fact made ordinary.

## 4.4 CULTURES OF LANDSCAPE: THE SELF, POWER AND DISCOURSE

### 4.4.1 Introduction

As was discussed at the start of this chapter, cultural geographies of landscape through the 1990s increasingly turned for inspiration and interpretative impetus to various forms of poststructural theory and critical cultural theory. In this section and the one that follows I want to trace out some of the substantive pathways forged by the adoption of these new critical and poststructural sensibilities. Here, first of all, I want to focus upon the work of some UK-based cultural geographers over the past ten to fifteen years, in particular the work of David Matless.

To begin with a rather general point, Don Mitchell's work on landscape might be thought of as one, particularly North American, response to new cultural geography's understanding of landscape as a way of seeing. As discussed in the previous section, Mitchell in a sense offers an at once broader and deeper version of the early cultural Marxist position of writers like Cosgrove and Daniels, extending their definition of landscape as visual ideology by detailing how landscape is also always simultaneously enmeshed in 'base' material processes of labour and production, in other words in everyday life and work. Here the influence of the work of a US-based generation of radical and materialist geographers such as Neil Smith and David Harvey is evident, as is a lingering, almost vestigial sense of a material (but not materialist) American landscape tradition including figures such as Carl Sauer and J.B. Jackson. In the UK, by contrast, it is probably fair to say that the clearest intellectual influence as regards studies of landscape through the 1990s was less variants of Marxist and radical thought and more conceptions of discourse, power and subjectivity devolving from the poststructural corpus of Michel Foucault. As the next section (4.5) will discuss, a reinvigorated analysis of geographical histories and geographical practices (e.g. Gregory, 1994; Clayton, 2000a; Driver, 2001) has been perhaps the most visible sign of this influence, but Foucault's thought is also to the fore when it comes to the specification and investigation of 'cultures of landscape', a phrase and an approach associated in particular at present with the work of David Matless (1992, 1993, 1995a, b, 1996, 1998, 2000, 2003) and a number of other current cultural and historical geographers (see, for example, Gruffudd, 1996; Linehan, 2003; Edensor, 2000; Lorimer, 2000; Merriman, 2005a, b). 'Cultures of landscape' might seem an opaque phrase, however it refers in one way quite straight-forwardly to everyday landscape practices such as walking, sightseeing, driving, boating and rock climbing, and then in a broader sense to the regulatory processes and cultural discourses through which notions of the proper conduct of such practices-in-landscape are elaborated.

Having given this initial definition it should be said that, with the exception of a relatively early paper on Foucault by Matless himself (1992), much work in this vein does not explicitly position itself vis-à-vis particular theoretical and epistemological approaches in the way that materialist or Marxist writing often does. In other words theoretical discussion is not really foregounded here, and, in a sense echoing Foucault's own genealogical and governmental enquiries, considerations of first-order or abstract questions regarding the status or definition of landscape are bypassed in favour of detailed discursive interpretation of particular figures, episodes and movements in cultural and intellectual UK landscapes histories. With this in mind, however, I do want to open this section by outlining some of the key motifs in Foucault's work, in particular his understandings of *discourse*, *power* and *subjectivity* as a means of clarifying the distinctiveness of the notions of landscape, self and culture at work here.

### 4.4.2 Michel Foucault: discourse, power and the subject

In his introductory essay to *The Foucault Reader* (1991), Paul Rabinow first seeks to clarify that Foucault's philosophy is strongly anti-essentialist in tenor. He helpfully puts this in procedural terms: instead of asking an abstract question such as 'is there an essential human nature?', Foucault will always aim to historicise and situate supposed abstractions, and so ask instead a question in the form, 'how has the concept of human nature functioned in Western society through history?' This historicisation of apparently universal or essential concepts like 'human nature' is a theme of much of Foucault's writing. His most notable achievement, perhaps, is to have shown that categories often assumed to be to universal, natural or objectively definable – categories such as sane/mad, healthy/ill, normal/deviant – are in fact historically and discursively constituted.

In Foucault's work, the word 'discourse', while retaining connotations of dialogue and speech, is taken to refer in an expanded sense to the totality of utterances, actions and events which *constitute* a given field or topic. And this definition alerts us to two things. First, a discourse is not just a set of written texts, an archive of documents. A discourse encompasses texts, speeches, dialogues, ways of thinking *and also* actions: bodily practices, habits, gestures and so on. Second (and this is perhaps the most crucial point), a discourse is not a series of things that are said about a separate, pre-existing entity – a landscape, say. A discourse of landscape is not a set of things said and done regarding a pre-existing, external and immutable 'landscape', already out there in the world. Instead a discourse of landscape *creates* landscape, makes it really, actually exist as a consequential and meaningful set of beliefs, attitudes and everyday practices and performances – and these collectively comprise what may be termed 'cultures of landscape'.

A discourse might thus be defined as a framework of intelligibility within whose bounds certain relationships, practices and subjectivities are, to use a phrase Foucault adopts from the painter Paul Klee, 'rendered visible'. As Barnes and Duncan (1992, p.8) note, 'discourses are both enabling and constraining . . . they set the bounds on what questions are considered relevant or even intelligible'. In other words, a discourse defines both what *can* and what *cannot* be said or done, what appears to be true, legitimate or meaningful and what is dismissed as false, deviant or nonsensical. Take the example of landscape again: a discourse of landscape, in a given setting and epoch, will enable and endorse some ways of behaving, perceiving, picturing and representing, but, at the same time, constrain or proscribe others. In this way, the political and ethical import of this notion of discourse begins to clarify: a discourse will establish some behaviours and identities as normal, approved and even natural, while making others appear unusual, marginal or unnatural.

At this point, clearly, the issue of *power* and its operations arises. For geographers such as Don Mitchell, writing in a broadly critical and materialist vein, questions of power – cultural, political, but above all economic – are questions of inequality, domination and resistance (e.g. Keith and Pile, 1998). Certain groups of people, institutions and ideas possess power, and so are dominant or hegemonic, are 'powerful' in the colloquial sense, while others are comparatively if not wholly powerless, and are pushed to the literal and figurative margins of the prevailing cultural, political and economic order. In this view, broadly speaking, power is concentrated in the hands of a minority, exercised over life rather than being part of life and, finally, negative rather than creative in its effects. Power is exercised to constrain, limit, forbid, detain and so on. It is crucial to note, however, that Foucault disagrees with this view, and advances instead an alternative, innovative account of power. He argues in particular that 'we must cease once and for all to describe the effects of power in negative terms. . . . In fact, power produces; it produces reality' (Foucault, 1977, p.155). Power, in other words, is productive, is what *creates* new subject positions and new regimes, new knowledges and practices. Moreover this productive power operates in a diverse and dispersed manner; it does not emanate from a single source. As Foucault (1981 [1976], p.93, emphasis added) notes – and this quotation is a key element of Matless's (1992) arguments on Foucault and landscape – 'power is everywhere, not because it embraces everything, but because it *comes from everywhere*'.

Here, Foucault is most definitely not seeking to argue away the palpable existence of inequalities, injustices and repressions of various forms. He is rather attempting to develop a more distinctive way of analysing how certain subjectivities and discursive formations acquire such concrete existence and persistence. And again at this point we can take landscape as an exemplar. It can be argued that certain forms of visual landscape (rural, picturesque, etc.) and certain forms of behaviour and practice in landscape (e.g. walking, painting) are, at different times and in different cultures, taken to be both aesthetically valuable and morally and physically uplifting. And this is not because some ineffable central 'power' coerces or dupes society into accepting such beliefs; rather it is a consequence of a much more anonymous and as it were horizontally distributed exercise of power; power exercised both over oneself and between and across selves. In a sense, therefore, in discursive terms, landscape may be first defined as a changeable matter of elaborated cultural conventions and ramifying codes of conduct (Matless, 1998, 2000).

One implication of this definition is that landscape norms and values are sustained by, and in an important sense simply *are*, a multitude of small, local, specific practices – 'cultures of landscape'. And here a fairly clear break from work which takes landscape to be a way of seeing, a visual ideology, is evident. Such a definition, as discussed in the previous chapter, sometimes understands landscape representations as a veil or curtain, a false or aestheticised vision obscuring a hidden reality. An important consequence is that a schism is therefore introduced between 'landscape-as-ideology' and landscape conceived in terms of practice, perception and lived experience, with the latter either being consigned to the realm of the personal and mystical, or else simply explained in terms of the outworking of ideological motifs and structures. This schism between ontological and ideological aspects is held by Daniels (1989) to be a key element in landscape's essential 'duplicity'. But Matless (1992, p.240) draws a slightly different point: 'the problem with such a division of analysis between the "ontological" and the "ideological" . . . lies not in possibly giving too much emphasis to one and not enough to the other, in somehow erring in the balance of ingredients, but in the polarity of the categories themselves'.

In other words a new understanding of the subject, and of everyday practice, is required, one which refuses to view these as either personal matters, standing somehow outside history and society, or else as the mere ciphers of ideological power. Here again, there is a turn towards Foucault's oeuvre, which indeed provides an alternative understanding of the subject, neither free and atomistic, nor an ideological fiction. First of all it must be noted that Foucault's early work, up to and including *The Order of Things* (1977), is often anti-humanist and anti-subjective. *The Order of Things* gradually moves towards an account of how, in the nineteenth century, emerging discursive formations and modes of knowledge, such as economics, linguistics and the life sciences, worked so as to establish new definitions of the human as both speaking subject and visible object. The self or subject – our modern sense of ourselves as free-standing individuals – is thereby presented largely as an effect of discourse.

Expressing these ideas in a different register, Foucault's more substantive and historical studies of mental hospitals (*Madness and Civilisation*, 1989 [1961]) and prisons (*Discipline and Punish*, 1991 [1975]) may be understood in terms of how curious, observant and disciplinary power/knowledge regimes in effect create subject positions such as 'the mad' and 'the criminal'. For Foucault, regulatory and disciplinary discourse, channelled and moulded via the emerging lineaments of the modern state, works so as to establish norms and to normalise human behaviours, most commonly through direct action upon the body – through procedures such as medical treatment, military training, confinement and visual observation, for example.

However, in Foucault's later work, from the late 1970s onward, a different account of the relationship between 'the subject' and discursive regimes is developed. As Harrer (2005, p.75) puts it:

> While in his earlier works, Foucault deals with topics such as 'Man's death' in the western 'epistêmê', or with mechanisms of how subjects are 'fabricated' and subjugated by disciplinary power, in his later works, beginning with the second volume of the *History of Sexuality*, Foucault develops an ethics that is based on a model of aesthetic self-fashioning. It is based on concepts such as 'aesthetics of existence', 'ethics of the self', or 'care of the self'.

Foucault (1982b, p.777) himself defines the task of his later works as 'a history of the different modes by which, in our culture, human beings are made into subjects'. The stress on different modes suggests that this 'making' of subjects is no longer only a matter of the operation of disciplinary power upon relatively benumbed and docile bodies, it is also just as much an exercise by the self on the self, a 'taking care' of the self, a self-fashioning, a cultivation of one's own body and mind through regimes of practice: reading, dieting, the analysis of one's dreams, sex and so on. It is important to note that this is not so much a 'turn' or epochal shift in perspective in Foucault's work as a transformative deepening of earlier insights through the addition of a new layer: the analysis of discursive regimes of power/knowledge is extended into the moral and ethical aspects of the art of living, of being oneself, and is thereby itself modulated. In this vein the second and third volumes of the *History of Sexuality: The Use of Pleasure* (Foucault, 1992 [1984]) and *The Care of the Self* (Foucault, 1992 [1986]) – focus upon practices of the self in the Ancient World. Thus also Foucault begins to write about 'technologies of the self' (see Martin et al., 1982), which 'permit individuals to effect by their own means or with the help of others a certain number of operations on their own bodies and souls, thoughts, conduct, and way of being, so as to transform themselves in order to attain a certain state of happiness, purity, wisdom, perfection, or immortality' (Foucault, 1982a, p. 18).

Key to this focus on 'inventing our selves' (Rose, 1996a) is the concept of *governmentality*, around which an academic literature has recently flourished (see Burchell et al., 1991; Barry et al., 1996; Rose, 1996a, b, 1999; and in geography, Braun, 2000; Barnett, 2001; Raco, 2003; Merriman, 2005a). Governmentality, or 'the conduct of conduct', is a term that describes both the governance of the conduct of entire populations (myriad policies and programmes concerning their health, education, feelings of citizenship and so on) and the manner in which individuals and groups conduct themselves, that is fashion themselves, through the elaboration of localised codes of conduct. Governmentality thus encompasses discursive practices across a wide ranges of fields and scales, practices aiming to regulate and 'improve' the behaviour of subjects through both formal and official codes, and through actions taken by subjects themselves. As Merriman (2005b, p.238) summarises, 'practices such as dieting, washing, walking, teaching, learning, and driving are all performed in relation to particular techniques and technologies of the self and (more broadly) technologies of government'. Governmentality describes, in other words, the conduct of the self from both without and within.

### 4.4.3 Landscape, action, conduct and citizenship

Collected together, these various Foucauldian ideas of power, subjectivity and government provide a lens through which the understanding of landscape proposed in particular by David Matless (see especially 1992, 1995a, 1998, 2000) may be pictured. As previously noted, however, Matless and others writing in this vein have not in general chosen to elaborate such ideas in any great detail (though see Merriman, 2005a, b) – hence their exposition over the past few pages. That said, much work in this area does emphasise the cultural and historical function of official, semi-official and regulatory discourses in supplying materials – textual, visual, corporeal – through which subjects might be fashioned, and might best fashion themselves, via specific landscape practices.

Such discursive forms are explored in depth in Matless's (1998) fullest statement thus far, *Landscape and Englishness*. As the title implies, the text focuses on connections between landscape, identity and citizenship in a specifically English, twentieth-century context, and in doing so it gathers together an eclectic and at times eccentric array of issues and practices: naturism, organicism, the Boy Scouts, town planning, campaigns against litter and ribbon development, ecology, fascism, socialism, health and fitness. This bricolage in turn hints at a distinctive interpretative methodology. In a review paper nearly contemporary with the production of *Landscape and Englishness*, Matless (1996, p.383, original emphasis), writes of a need for what he terms 'melded materiality and semiosis', and thus for cultural geographies 'operating less through images of landscape, than the tracing of objects and practices (which may include paintings, writings etc.) *producing* landscapes'. As earlier noted, the key point here is that landscape is a cultural production composed of actions, beliefs and practices, and not a stage on which practices are played out. A fairly clear distance is thus established between Matless's conception of landscape interpretation and his immediate geographical predecessors such as Denis Cosgrove and Stephen Daniels. This is cemented by the invocation, early in *Landscape and Englishness*, of Foucault's genealogical historiography. Genealogy, as defined by Foucault (1984, p.59), is:

> a form of history which can account for the constitution of knowledges, discourses, domains of objects etc., without having to make reference to a subject which is either transcendental in relation

to the field of events or runs in its empty sameness throughout the course of history.

Matless takes first from this a distinct sense of the contingency and plurality of history, running counter to any teleological tale of evolution or progress in human affairs. A more specific injunction is that neither 'landscape' nor 'Englishness' should possess any essential or inherent qualities; in their 'unstable heterogeneity' (Matless, 1998, p.21) they rather exist forever as versions and variations without an original. In consequence (and here the indebtedness to Foucault is clear),

> the question of what landscape 'is' or 'means' can always be subsumed in the question of how it works; as a vehicle of social and self identity, as a site for the claiming of a cultural authority, as a generator of profit, as a space for different kinds of living.
>
> (Matless, 1998, p.12)

Investigating how landscape works, for Matless, involves attending to both the highways and byways of cultural discourse, to official state policies and well-remembered figures and events, but also to more marginal movements and 'arts of living' whose precepts now seem arcane or absurd. Substantively, in *Landscape and Englishness*, the focus is upon debates over the English countryside and English citizenship through the mid-twentieth century. Following Foucault's lead, Matless focuses upon English bodies, and the panoply of discursive practices seeking to regulate and channel bodily actions in ways deemed most apt for the production of modern, 'landscaped', English citizen-subjects. Hence we see, in particular between the wars, various and varied campaigns clustering around the English body hygiene, health and fitness (Matless, 1995b), organic produce, rambling and 'proper' knowledge of the countryside as gleaned through correct visualisation, and the deployment of practices such as map and compass reading. At the same time, debates churn over the form of the English landscape, impelled by processes such as suburbanisation, increased car use (leading to greater leisure access), the erection of electricity pylons, ribbon development and so on. In bringing these matters together under the rubric of 'cultures of landscape' one of Matless's achievements is to link landscape, subjectivity and citizenship indelibly together, thus revealing both the production of new subjectivities via landscape practices and the concomitant production and mobilisation of landscape, both material and symbolic, within discourses of health, identity, authenticity and citizenship. In this way, the central aim is, 'to consider the production of the bodily subject through landscape, where that landscape is not simply a backdrop to action but a culturally charged object. . . . Landscape becomes the subject of codes of conduct and aesthetics of existence' (Matless, 2000, p.142).

Another notable point here is that, in focusing explicitly upon twentieth-century histories, Matless develops what might be termed a new ambit and resonance for the term landscape, quite outside of its poststructural re-theorisation. What I mean by this is that, first, the move away from a pictorial and 'art historical' understanding of landscape effects a substantive shift away from the latter eighteenth- and nineteenth-century contexts – landscape art's cultural zenith. And so the intellectual centre of gravity of landscape studies moves from *property* to *propriety*, that is from landscape understood as an artistic form in the service of an elite, country-estate vision of land, culture and society to landscape thought of as a matter of conduct and forms of 'proper' bodily display and performance.

Second, in terms of ambit and resonance, Matless's writing can be read in part as an attempt to rescue the very word landscape from its popular cultural consonance with notions of prettified heritage and melancholy nostalgia for a mythical rural past (see Hewison, 1987). As was discussed in Chapter 2, he reads the work of the noted landscape historian W.G. Hoskins as emblematic of a growing post-war intellectual refusal of English modernities, and a symbolic retreat instead into a nostalgic vision of pre-industrial and local landscape value (Matless, 1993). In consequence it is unremarkable for a contemporary critic such as Cresswell (2003, p.269) to complain that landscape studies are 'too much about the already accomplished', and that the word itself is 'altogether too quaint'. But the point here is that our association of the word landscape with the pre-modern, the pretty and the pristine is actually a relatively recent cultural phenomenon. Thus, in *Landscape and Englishness* Matless demonstrates at length how notions of landscape, the body and identity have, in fact, been consistently enrolled in the service of competing visions of English modernity. In particular, the text excavates what it terms a 'planner-preservationist' vision of a modern English landscape, one alive between the wars, and, for a time in the post-war period, ascendant in policy terms – as is visible in UK post-war reconstruction projects, Green Belts, 'new towns', motorway systems, National Parks, and the entire regulatory and academic edifice of 'Town and Country Planning'.

For planner-preservationists such as Patrick Abercrombie and Dudley Stamp, a newly landscaped English modernity would be informed and guided by principles of rationality, design and order, and would be effected by an enlightened vanguard of experts. The important point is that this constitutes a particular *version* of what modernity might be; as Matless notes, 'an orderly, serious and concrete modernism is embraced as opposed to a "jazz" modern which is put down as frivolous, even ornamental' (1998, p.51). This particular modernity was therefore quite in hock to notions of direct State planning and landscape control, as exemplified by the autobahns of 1930s Germany, while at the same time wishing to preserve what it saw as the proper forms of landscape and Englishness (neat fields, well-nucleated towns and villages, physically upright citizens) against the encroaching forces of a ramshackle and bastardised Americanism (unchecked ribbon growth, suburbs, jazz and uncontrolled dancing in the fields). As this shows, 'different political aesthetics work through different senses of order in landscape. Preservationists offered a scene of leadership and action, with the expert rather than the ordinary person the key shaper of the land' (*ibid.*, p.30).

The wider point for Matless is that these visions of self and landscape are coloured by – in fact *are* – sets of moral and aesthetic values. Landscape is an issue of propriety and thence more generally morality. This confluence of moral and landscape values has been examined further by Matless in the specific context of leisure practices in the Norfolk Broads (Matless, 2000), and it forms a central thread of other work developing a 'cultures of landscape' perspective (for example, Brace, 2000; Edensor, 2000). In terms of moral bodily conduct, for the 'planner-preservationists', whom Matless (1998) focuses on, the inculcation of a formal 'art of living' was as imperative as discovering 'the means of correct training' (Foucault, 1991b [1975]) of prisoners had been for penal innovators in the eighteenth century. Thus, 'For preservationists walking, cycling, camping and map-reading made up an "art of right living" whereby individual and nation might give form to itself environmentally, generating intellectual, moral, physical and spiritual health' (Matless, 1998, p.62). In turn, however,

> arguments for citizenship always worked in relation to a sense of 'anti-citizenship'. While a landscaped citizenship is set up as

potentially open to all and nationally inclusive, it depends for its self-definition on a vulgar other, an anti-citizen whose conduct, if not open to re-education, makes exclusion necessary.

(Matless, 1998, p. 62)

Matless paints a characteristic canvas of anti-citizenry: 'Cockneys', jazz afficionados, reckless boaters. More is at stake here, though, than abstract distinctions reflecting class divisions and outright paternalism and elitism. The intertwining of landscape, body and nation is effected not just symbolically but materially and performatively, at the precise level of the body-in-landscape:

landscaped citizenship worked through a mutual constitution of the aesthetic and the social, the eye and the body. The aim of extending visual pleasure to the people was tempered by a desire to control potentially disruptive bodily effects. The education of the eye was to be accompanied by a self control in the body.

(Matless, 1998, p.63)

Matless goes on to explore in rich and at times lugubrious detail how such arguments were played out in contexts including motoring, scouting, mass rambling and naturism. He shows how the landscape discourse of the planner-preservationists rubbed up against other distinctively English modernities – the emerging ethos of 'daytrippers' and suburbanites, the strangely right-wing and rustic 'organic' movement of the 1930s and the better-remembered northern English socialist campaigns for open access to the countryside, as exemplified by the Kinder Scout Mass Trespass of 1932 (see Darby, 2000). I want to conclude this section, however, with some more general remarks. The first, with an eye to the fact that the work of Matless and others in this area is characterised above all by archival richness and texture, would be that in some ways what we see here is a stance characterised more by an attentiveness to the fine grain of cultural histories than adherence to any theoretical or epistemological credo regarding the status of 'landscape'. This is cultural history in some ways, and so it very much develops the agenda of the first generation of 'new' cultural geographers by further cementing the links between geographical writing and the interpretative approaches of the arts and humanities. In a similar vein, it would be invidious to suggest that Matless's work somehow simply rejects the insights garnered by understanding landscape as a way of seeing or visual ideology. As this section has shown, quite different notions of power and subjectivity inform landscape studies here, and landscape itself moves from 'image' to 'practice' in a distinctive fashion. But equally Matless has himself written on the links between landscape, vision, power and authority, and one of the central themes of *Landscape and Englishness* is attempts by different groups to claim discursive authority over landscape.

Most recently Matless (2003) has chosen to proffer a description of landscape as a 'shuttle', weaving together disparate realms of economic, political, scientific and aesthetic practice, and this does seem an apt metaphor for concisely expressing the understanding of landscape advanced here. Thought thus, 'landscape can be considered a term which productively migrates through regimes of value sometimes held apart' (p.231); landscape enables the critic to weave together narratives of morality, perception, bureaucracy, advertising and so on. What we have then, is a productively diffracted vision of landscape, 'always already natural and cultural, deep and superficial . . . impossible to place on either side of a dualism of nature and culture, shuttling between fields of reference' (ibid., p.231). This way, we can come across incongruous but telling images:

> In the downstairs gallery, a 'Litter can Kill' exhibit included a deflated 'Mr Blobby', the pink-and-yellow spotted sidekick of TV celebrity Noel Edmonds, and at the time something of a cult among children and adults alike. The binning of Blobby was somehow symbolic of all the Broads Authority had been trying to achieve . . . Blobby lay, less than potent, among cans and fishing line and hooks and mess; a plastic thing out of place, something that should never have come to the Broads. Children and adults alike should take such rubbish home.
>
> (Matless, 2000, p.162)

## 4.5 LANDSCAPE, TRAVEL AND IMPERIALISM

### 4.5.1 'Imperial landscape' and postcolonial theory

One writer quoted approvingly by Matless (1998) is the visual theorist W.J.T. Mitchell (2002 [1994b]), editor of an influential collection of essays on *Landscape and Power*. In his short introduction to this collection, and in his chapter on 'Imperial Landscape' which immediately follows, Mitchell sets out a series of significant claims and agendas regarding the study and status of landscape. First, he writes, 'the aim of this book is to change "landscape" from a noun to a verb. It asks that we think of landscape, not as an object to be seen or a text to be read, but as a process by which social and subjective identities are formed' (Mitchell, 1994b, p.1).

To think of landscape in verbal terms ("to landscape") involves phrasing landscape in terms of action, activity and performance: landscape is a doing, to landscape is so to *do* something, landscape is therefore not just an inert 'thing', an object to be viewed or a neutral background. Mitchell argues that this perspective will at once transcend and encompass previous art historical paradigms of landscape study; both the traditional, aesthetic contemplation of landscape imagery and contextual, iconographic and historical readings of landscape symbolism (in some ways similar to the 'iconography of landscape' proposed by Cosgrove and Daniels (1988)). In this way:

> *Landscape and Power* aims to absorb these approaches into a more comprehensive model that would ask not just what landscape 'is' or 'means' but what it *does*, how it works as a cultural practice. Landscape, we suggest, doesn't merely signify or symbolise power relations; it is an instrument of cultural power, perhaps even an agent of power.
>
> (Mitchell, 1994b, p.1, original emphasis)

Adherents of both materialist and iconographic approaches to landscape might wish to query the distinction implied here between 'merely signifying' and acting as 'an agent of power', given that so much work in these areas has sought to show how the act of representation (for instance in depictions of landscape) is inextricably linked to the exercise of power in various forms. But Mitchell also wants to stress this point, arguing further that an account of landscape as an active agent 'has to trace the process by which landscape effaces its own readability and naturalises itself' (*ibid.*). Thus we encounter again the notion, already familiar from accounts of landscape in terms of property (Berger, 1972; Williams, 1985; Daniels, 1989), perspective (Cosgrove, 1985) and ideology (Mitchell, 1998), that landscape is a gaze upon or representation of the world which masquerades as 'objective', 'authoritative', 'accurate', 'realistic' and so on, while in actuality being a culturally specific artifice. The distinctiveness of Mitchell's argument perhaps lies in the degree to which landscape is understood in terms of *process* and *movement*.

> landscape is a dynamic medium, in which we 'live and move and have our being', but also a medium that is itself in motion from one place or time to another. In contrast to the usual treatment of landscape aesthetics in terms of fixed genres (sublime, beautiful, picturesque, pastoral), fixed media (literature, painting, photography), or fixed places treated as objects for visual contemplation or interpretation, the essays in this collection examine the way landscape *circulates* as a medium of exchange, a site of visual appropriation, a focus for the formation of identity.
>
> (Mitchell, 1994b, p.1, original emphasis)

In one sense Mitchell is simply advocating a newly interdisciplinary approach to landscape, with insights from visual and critical theory cutting across the traditional domains of art history, English literature and so on, in order to capture a richer, more mobile sense of landscape. At the same time, the central idea here is that landscape *travels*: not just that material landscapes might be literally transported, but that the values, beliefs and attitudes that work through and emerge from specific landscape practices and 'ways of seeing' can be seen to migrate through spaces and times. And this idea takes centre-stage when Mitchell comes to his own substantive chapter (as opposed to editorial overview). Here, he writes that '*landscape is a particular historical formation associated with European imperialism*' (1994a, p.8, my emphasis).

In this, the final substantive section of this chapter, I want to discuss work which, over the last ten to fifteen years, has traced out a series of linkages between landscape, travel and imperialism. Cultural and historical geographers have been notably to the fore here, in a sequence of texts by authors representing something of a 'second generation' of culturally and theoretically informed writing (e.g. Driver, 2001; Clayton, 2000a; Ryan, 1997; Phillips, 1997; Blunt and McEwan, 2002) post the 'cultural turn' in human geography. In some ways this body of work is the most significant outworking of the reorientation of landscape studies achieved by authors such as Cosgrove, Daniels and Rose, and it is worth noting that several of these authors have themselves moved to write on issues of 'imperial land-scape' and travel (e.g. Cosgrove, 1999; Duncan and Gregory, 1999). As

I will discuss in more detail below, we can also see in these and other texts something of a segue towards more poststructurally inclined landscape geographies. But in a more basic sense the connection between landscape, travel and imperialism involves extending the analyses previously applied to, for example, British or Italian contexts, to non-European spaces, and thereby situating landscape within a more extended and complex terrain. This is precisely Mitchell's (1994a, p.9) argument: 'we need to explore the possibility that the representation of landscape is not only a matter of internal politics and national or class ideology, but also an international, global phenomenon, intimately bound up with discourses of imperialism'.

While imperialism is usually taken to refer to explicit political, economic and military projects aimed at conquest and domination, much critical writing in the arts and humanities has also discussed imperialism as a way of thinking, a complex set of attitudes which, through their expression and reproduction in art, literature, science, academic writing, media and so on, work so as to perpetuate European and latterly Western perceptions of superiority over other cultures, belief in their right to govern, and faith that they are in possession of essential truths and insights. In other words, to make what is a rather blunt initial statement, much European and Western scientific, intellectual and artistic practice may be understood as in some way contributing to and reflecting imperialist ideologies and discourses. And thus, alongside many other cultural forms, the artistic, literary and scientific representation of non-European landscapes by Europeans – by travellers and explorers, for example, and by officials, the military, colonists and settlers, traders and pilgrims – can be critically examined as part of more general 'discourses of imperialism'. It is important to stress, however, that this is not a case of certain landscapes somehow simply being 'imperialist'. As Mitchell elaborates, landscape

is not to be understood . . . as a mere tool of nefarious imperial designs, nor as uniquely caused by imperialism. . . . *Landscape might be seen more profitably as something like the 'dreamwork' of imperialism*, unfolding its own movement in time and space from a central point of origin and folding back on itself to disclose both utopian fantasies of the perfected imperial prospect and fractured images of unresolved ambivalence and unsuppressed resistance. In short, the posing of a relation between landscape and imperialism is

not offered here as a deductive model that can settle the meaning of
either term, but as a provocation to an inquiry.

(Mitchell, 1994a, p.10, my emphasis)

Many of the key elements of cultural geographical enquiry into relations
between landscape and imperialism can be distilled from the ideas
expressed in this citation. In the first place there is a general acceptance of
landscape's uniquely European character. As Mitchell says, its 'central point
of origin' as an artistic genre, a mode of representation and a particular
way of seeing and perceiving the world is accepted to be the development
of new understandings of perspective, perception and subjectivity in the
Italian Renaissance (Cosgrove, 1985). And standard art histories of landscape describe, *inside* Europe, the evolution of distinctive national tastes
in landscape: classical seventeenth-century Dutch landscape, the British
picturesque in the late eighteenth and nineteenth centuries, German
nineteenth-century Romantic landscape, and so on. Second, enquiry here has
focused upon *how landscape travels*, and the implications of this movement for
landscape, for senses of identity, and for understandings of the meaning and
purpose of travel itself. Thirdly, in terms of 'utopian fantasies of the perfected
imperial prospect', a key concern for many cultural geographers has been
to demonstrate how certain ideal European landscape forms have been used
to characterise, appropriate and judge non-European scenes. Both classical
aesthetic notions of order and proportion in landscape and also genres of
landscape aesthetics such as the picturesque, the pastoral and the sublime
were used by European travellers and colonists as means of understanding,
evaluating, inhabiting and *making knowable* non-European landscapes. And
then, finally, turning to 'fractured images of unresolved ambivalence and
unsuppressed resistance', many writers have sought to stress that imperialist
desires to simply export, project and impose the epistemological and
aesthetic values associated with European norms of gazing and representing
were never in fact absolutely realised – in other words, the colonial or
imperial situation is one which is always already complex, a fractured and
hybrid 'contact zone' (Pratt, 1992), wherein local and indigenous knowledges and practices are incessantly braiding and disrupting European
imperial visions.

Before moving on to discuss examples of these intellectual agendas
concerning landscape and imperialism, it is important to note that they gain
much of their impetus, direction and critical politics from the emergence

of postcolonial writings on colonialism and imperialism. In turn, contemporary postcolonial theories are heavily indebted to various strands of poststructuralism (for an overview see Ashcroft *et al.*, 1995). For example, the approach taken by one of the founding texts of postcolonial studies, Edward Said's (1978) path-breaking *Orientalism*, is inspired by Foucault's work on discourse, power and representation. As discussed in the previous section, Foucault demonstrates how various discursive power/knowledge regimes work as dividing practices, establishing distinctions between the sane and the mad, the law-abiding and the criminal, the healthy and the ill – in general, distinctions between a normalised 'self' and a marginalised and deviant 'other'. Said's work applies these insights and procedures to another notable distinction: that made between the West and the Orient. When, using Foucault's discursive analyses, we consider the academic and foreign policy fields of 'Oriental studies', and more broadly Western artistic and literary representations of 'the Orient', what we begin to see is that 'Orientalism' in *toto* comprises a discursive field that produces and perpetuates certain images of the Orient in relation to 'the West'. A Western self is thus produced alongside an Oriental other. And Said's argument is that this process of subject-formation is decidedly *imperialist* in character, in so far as it associates the West with qualities such as rationality, authority and self-control, and the Orient, by contrast, with degeneracy, backwardness and impulsiveness.

For Said (1993), therefore, orientalism is an 'imaginative geography', a sort of moral mapping of the world in which non-Western cultures and lands are described through negative contrasts with a pre-given and pre-eminent Western self. And numerous geographers have further pursued Said's adoption of Foucault as a critical historian of Western thought and practice, in particular scrutinising the discursive practices of exploration (Driver, 2001), travel (Duncan and Gregory, 1999) and geography itself as an intellectual tradition (Gregory, 1994, 1995). At the same time they have further delved into the nexus of postcolonial and poststructural theory by using Jacques Derrida's (1976, 1978) early work on deconstruction and the 'metaphysics of presence' to examine the (de)stabilisation of binaries such as West/East and Self/Other (e.g. Blunt, 1994; Barnett, 1998). Derrida's writing offers clear purchase for the projects of postcolonial criticism because it is, in part, an attempt to disturb, unsettle and think otherwise various claims to presence, identity, authority, originality and truth. For Derrida, much Western art, literature and philosophy can be understood in the context of claims on these lines. However, he argues, such claims are always already unstable, because the definition and expulsion of what is *not* truth or presence will incessantly return to haunt the supposed original – will be, in fact, its 'constitutive outside'. Thus a definition of presence will rely upon absence for its meaning, a definition of 'Western' will always be circumscribed by the likewise definition of the 'non-Western', and so on. And thus a couplet such as Self/Other can only ever describe a shifting territory of meaning. As a consequence, any textual or visual representation *of* the Other *by* the Self will, if examined with sufficient acuity, be shown to undercut its own attempts to divide, purify and presence. This insight is thus especially productive as regards the analysis of how indigenous or native voices are variously silenced and erased by forms of imperial discourse such as landscape representation. In turn, the recuperation of alternative, overlooked and marginalised perspectives, and the decentring of Western epistemologies and histories, become key agendas for postcolonial studies, for example as articulated via subaltern studies (Spivak, 1988), and calls to 'provincialise Europe' (Chakrabarty, 1992).

Moving on to issues of landscape specifically, three broad strands of postcolonial work utilising these Foucauldian and Derridian insights can be identified: writing on landscape as an objective and scientific gaze, on landscape as a distinctively European set of aesthetic values and on landscape as itself an impetus for forms of travel disclosing new models of subjectivity and perception.

### 4.5.2 Science, observation and authority

First of all, recalling the landscape way of seeing's close association with Western sciences of observation and classification, it becomes possible to speak of landscape in non-Western contexts as an 'objective', 'scientific' and thereby peculiarly *imperial gaze*. It is important to remember here that landscape is being defined as a particular mode of looking and representing, and thus that when we speak of landscape we are referring to the gaze of a particular subject or self. This person is specified in Mary Louise Pratt's (1992) seminal *Imperial Eyes*: he is 'the European male subject of European landscape discourse – he whose imperial eyes passively look out and possess' (Pratt, 1992, p.7). For Pratt and for a series of others exploring similar ideas (e.g. Blunt, 1994; Ryan, 1996; Naylor and Jones, 1997; Phillips,

1997), the landscape gaze, projected out through the eyes of European explorers and scientists from the seventeenth to the twentieth centuries, is quite often a detached gaze, a controlling gaze, a gaze that somehow leeches the life out of the scene surveyed and replaces it with either a fabricated set of Eurocentric preconceptions or a *tabula rasa*, an emptiness, blank but measurable. There are several conjoined elements to this analysis. In the first place the landscape gaze here connotes height and command, it is an elevated prospect, from which position observers are, so to speak, 'masters of all they survey'. The commanding prospect, offering objective, authoritative and wide-ranging vision, and establishing the viewer in a place of epistemological and juridical supremacy, is a classic trope within the art and literature of imperial travel and exploration, and it has been examined in varying contexts by a number of cultural geographers (for example, Kearns, 1997, on the traveller Mary Kingsley; Cosgrove, 1994, on the Earth seen from space; Barnett, 1998, on nineteenth-century travel in the African interior). Further, and relatedly, as scientific gaze and prospect, landscape is once again linked quite closely here to the 'practical sciences' of cartography and navigation: Clayton's (2000b, 2000c) in-depth analysis of the visual and cartographic practices on board George Vancouver's late eighteenth-century voyages to the Pacific north-west provides a good example of how landscape gazes, map-making and navigation intertwine in practice. In this sense, landscape clearly also functions as a particular vehicle through which the unknown is rendered visible and legible; as Ryan's (1996) study of explorers in the Australian interior makes clear, landscape imagery can be enrolled alongside the sureties and certainties of mathematics and geometry as a means of assuring clear and distinct topographic knowledge and command. And, finally, in tandem further with the pictorial and classificatory impulses of the emerging life sciences in the nineteenth century, landscape representations, based on principles of proportion and perspective, present themselves with a stamp of accuracy, reliability and trustworthiness (see MacKenzie, 1990; Pratt, 1992; Driver and Martins, 2005). Images of landscape take their place alongside more microscopic scientific illustrations of flora and fauna, and other visual forms such as navigational charts and sketches, as part of an overall system of faithful and repeatable observation. In sum, there are a complex series of linkages between landscape as a mode of perceiving and picturing and the growth and expansion of European geographical and scientific knowledges under imperialism. Landscape images and descriptions in this way became part of an increasingly standardised discursive repertoire through which non-European and 'far-flung' spaces were made visible and understandable for then emerging European scientific establishments, imperial bureaucracies and broader metropolitan publics.

---

**Box 4.2 CAPE EVANS, ROSS ISLAND, ANTARCTICA***

Throughout the eighteenth and nineteenth centuries, European exploration of other continents, and of the poles and tropics, was an important conduit through which senses of landscape, identity and difference were produced, confirmed and expressed. European exploration of the globe, and subsequent settlement and colonisation, produced landscapes both material and symbolic. It also served as a context through which distinctive *cultures of landscape* – certain ways of perceiving landscape, bodily attitudes and postures, and semi-formal if often-unspoken codes of conduct for behaving in landscape – could develop.

In January 1911 a British Antarctic expedition of around thirty men led by Captain Robert Falcon Scott landed on Ross Island, on the edge of the continent, and made their base at a spot they named Cape Evans. The first aim of this expedition was to reach the South Pole, but the majority of their time was spent at home at Cape Evans, as it were, hibernating through the winter darkness, preparing clothes and equipment, conducting scientific surveys and investigations, and becoming more familiar with the local Antarctic environment.

The hut built by the British expedition at Cape Evans opened out onto contrasts and possibilities: local coastlines and coves, small islands out in the frozen waters of the bay, glaciers and ice slopes behind in the foothills of Mt Erebus. This was a space for observing

---

* This box draws upon material previously published in: Wylie, J. (2002) 'Earthly poles: the Antarctic voyages of Scott and Amundsen', in Blunt, A. and McEwan, C. (eds) *Postcolonial Geographies*, London: Continuum, pp.169–184.

and moving through and playing in – in other words a space for science, art and leisure. And the ways in which the British expedition engaged with these various activities demonstrates how culturally specific ways of seeing and inhabiting landscape are forged and performed.

First of all, Cape Evans was a landscape to walk in. Throughout the nineteenth and early twentieth centuries in the UK, largely via the elaboration and popularisation of modes of observing and practising landscape such as the picturesque and the romantic, walking in the 'great outdoors' became a significant way in which aesthetic and moral values were expressed (see Andrews, 1984; Wallace, 1993; Matless, 1998). That is to say, walking was understood to enable deeper and closer appreciation of natural scenery, and, as a physical, visual and educational activity, it was seem as a way of bettering oneself, of becoming a physically and morally healthier person. For the upper and emerging English middle classes, walking-in-nature denoted taste, education, companionability and a set of shared understandings and behaviours regarding the value and conduct of outdoor activity. This is why, in 1911, and on the then almost completely novel terrain of Antarctica, Edward Wilson, the scientific leader of the British polar expedition, could rebuke 'one or two who will stay indoors too much – which is a very great mistake' (Wilson, 1972, p.77).

Walking in the icy environs of Cape Evans was thus a practice through which Captain Scott and his party could literally make Antarctica visible and knowable as a cultural landscape composed of spiritually and aesthetically healthy activities. Getting out and about was the way to know both Antarctica and yourself better. Of course this sometimes led to mishaps, chronicled with mounting exasperation by Scott in his diary, as his men variously wandered off in blizzards, went on ill-advised cycling excursions, and fell off icebergs. But at no stage were these activities curtailed. In contrast, Scott (1913, p.207) regarded it as significant that there was 'no-one who has the least prospect of idleness'. Improving strolling therefore both *made* and *claimed* the Cape Evans landscape as

continued

English, in a way: even on arrival Scott was already noting that its 'vast tracts of rocks for walks' might engender 'a homely feeling' (*ibid.*, p.65).

Behind a culturally ingrained belief in the value of outdoor walking – something common, shared and understood amongst educated Edwardian Britons such as those on Scott's polar expedition – more subtle codes and practices channelled the relationship between self and landscape around Cape Evans. In his classic memoir *The Worst Journey in the World*, Apsley Cherry-Garrad, a member of the expedition, described his companions as 'artistic Christians' (1939, p.i). In this specific place and time, this odd phrase in fact described what *scientists* should be, in particular it described the ideal spirit in which scientific exploration of the natural world should be conducted. Many of the explorers, chief among them the expedition's scientific leader Edward Wilson, held deep religious convictions. For Wilson and his followers, science *was* art in so far as accurate, factual and empirical observation held intrinsic aesthetic value. Science was furthermore a marriage of the human mind and natural world which revealed a divine order in things. A common feeling on the British polar expedition was one of having been specially selected to witness privileged and exceptional scenes, and the Antarctic landscape was correspondingly understood to be something fateful and providential. In this way Captain Scott's scientists observed landscape as they observed Sunday service, and they so made Antarctica visible as a terrain for a robust Anglican empiricism.

These unfolding relations at Cape Evans illustrate how landscapes may be understood to function as evolving concordances of mental and material topographies, as productive combinations of thought, action and environment. To put this another way, it shows how landscape may be understood in terms of *discourses*. Discourses are sets of beliefs, knowledges and practices that both enable and constrain what can be said and done with respect to any given subject. Thus a particular culture of landscape, such as that of walking and viewing in the Edwardian Antarctic, is a discourse which sets a limit on what constitutes acceptable forms of walking, and, within that limit, enables the expression and elaboration of new and innovative cultural forms of walking.

### 4.5.3 Landscape aesthetics and the representation of the non-European

A second major strand of analysis pertaining to landscape, travel and imperialism has focused less upon questions of authority and objectivity within the landscape gaze, and more upon the role of European landscape aesthetics in representing non-European scenes, although it is important to note that cultural geographers and historians have been concerned to demonstrate that these two elements cannot be wholly separated in practice, given the way in which aesthetic principles regarding landscape form and composition are clearly intertwined with social status, visual authority and 'correct' modes of gazing (for example, see Hevly's (1996) analysis of the conjoined scientific and aesthetic currents of nineteenth-century glaciology). That said, several critical arguments have been made with specific reference to landscape form and aesthetics in the context of exploration, colonialism and imperialism, and the first of these is that the very visual structure of conventional landscape art has the effect of *subduing strangeness*, of making the faraway and the topographically alien familiar to European eyes. In other words, as W.J.T. Mitchell (1994a) himself notes, elements such as the perspectival vanishing point, towards which the space of the picture converges, the *repoussir*, or side screen, acting as an internal shelter and frame, and the serpentine 'line of beauty', guiding the gaze off into the middle distance of the scene – all these various devices and conventions act so as to render non-European scenes in distinctively European fashion, and thus may be understood as part of a more general process whereby the world is made visible, legible and knowable. Of course, this familiarisation can be understood as a sort of taming, preserving a sense of difference, but only within a known idiom. In this vein, for example, Duncan (1999) provides an intriguing account of the dissonance produced when a far-flung landscape (here Sri Lanka) actually did appear to resemble 'home'. And sometimes, alternatively, the codes and clarities of landscape representation might founder in the face of species and topographies altogether alien to European vision, for example on the South American coastline (Martins, 1999), or on the anti-perspectival mirror of the Antarctic interior (Pyne, 1988; Wylie, 2002b, c). But even here, as Pratt (1992) demonstrates, landscape art could still provide a visual stage upon which strange or inexplicable phenomena might be displayed and discussed.

These points lead us into the important question of the visual and textual representation of non-European peoples within imperial landscape. As was discussed in Chapter 3, an early agenda for critical art historians such as Bermingham (1986) and Barrell (1980) was to show how the conventions of picturesque landscape worked so as to symbolically legitimate forms of private landownership, and, at the same time, depict rural labourers as timeless, distant 'objects' in landscape. Thus 'the rural poor' could be envisioned as simply part of the scenery, so to speak, or in extremis as elements only adding visual charm and interest; certainly in this way their status as active agents involved in the transformation and making of landscape could be effaced. Subsequently, many critics of colonial and imperial landscape, such as Pratt (1992), Outram (1995), Birtles (1997), Wheeler (1999) and Ryan (1996), have sought to show how a similar process played out in colonial 'contact zones' as diverse as Polynesia, Africa and Latin America. Beyond being represented as exotic objects of curiosity, or as 'natural features' of the landscape alongside diverse flora and fauna, indigenous non-European peoples were, again, familiarised. That is, they were rendered through visual idioms already known from European landscape art, the most common of these, as Outram (1995) discusses, being the Arcadian shepherds and nymphs of classical mythology, so common in the Italian and French landscapes of Claude, Poussin and other popularisers of the genre.

Such rendering, it has been argued, paints the colonial situation in soft focus, lifting it out of historical actuality into an aestheticised realm of Edenic abundance, innocence and harmony. European visions of the faraway, and in particular visions of Australasia and the Pacific, persistently invoked the image of a pre-lapsarian, pre-social 'Golden Age', found again. Such images are voyeuristic in so far as the exoticised and naturalised non-European other becomes the object of sexual curiosity and the subject of imputed sexual appetite and licence. And they are equally narcissistic in their yoking together of the 'faraway' and the 'long ago', their implicit claim thus being that the geographical distant is equivalent to the historically distant. In other words, non-Europeans resemble how Europeans used to be, in the past, at a less advanced, more primitive stage in their development. Thus these landscapes arguably adhere to 'a coloniser's model of the world' (Blaut, 1993), one in which Europe is portrayed as the fulcrum and motive force of human histories.

One important consequence of this naturalising and aestheticising of the non-European 'other' is that non-European *landscape* is equally simultaneously pictured as natural and pristine, as untouched and untransformed. This symbolic erasure of other possible histories of land occupation of course parallels more literal processes of imperialist land appropriation and indigenous dispossession, for example in North America and Australia. It also, as Ryan's (1996) analysis of explorers in Australia and Neumann's (1996) work on early African nature preservation argue, tends to 'empty' the landscape, just as much as cartography advances a blank space of the unknown before itself. In this way, as untouched nature, the landscape is pictured as ripe for settlement and colonisation: the landscape way of seeing, these writers argue, is incipiently colonial. Further, and lastly with regard to issues of aesthetic depiction, the picturesque and pastoral idioms so commonplace in the visual and textual landscape descriptions of explorers, traders and colonists extended this vision of a far-off idyll through evaluating non-European landscapes in terms of their *fruitfulness*. In the colonial and imperial context the landscape way of seeing chimes with an emerging capitalist understanding of nature as a *resource*. Landscape thus apparently pictures itself as already ready for improvement and transformation, for crop and pasture.

### 4.5.4 Landscape, travel and the self

Beyond the issues of control, authority and appropriation raised by analyses of the scientific and aesthetic dimensions of landscape representation in non-European contexts, a third and final theme of writing in this area has involved considering how notions of landscape, visual culture and travel entwine to produce new cultural ideals and senses of self. In the plainest terms it can be argued that landscape *provokes* travel: a thirst for particular types of scenery, and a curiosity regarding the antique and exotic, in part impels the discursive construction of purposeful voyaging as a cultural ethos. The specific argument here is that in different ways, different forms of movement – the poet wandering in a nature newly realised as sublime, the explorer charting new terrain, the fieldworker enumerating empirical curiosities, the nascent tourist in search of exotic sights – render the world as an object to be visually consumed. Western cultures of travel, tourism and exploration have the general effect of producing the world as scenery, spectacle – in this sense as *landscape* (Urry, 1990). In tandem with this process, the act of travelling has increasingly been understood as an act of self-improvement, self-discovery and ultimately self-definition. As Chard (1999) notes, within diverse Western modernities travel has been understood as a movement of the self into the other, a movement in which both self and other take shape and definition. Cultural geographers and cultural historians have in consequence explored many specific instances of this general process, notable examples being the eighteenth-century 'Grand Tour' (Chard, 1999; Calaresu, 1999) and, slightly later, the British cult of picturesque and sublime landscape (Andrews, 1989; Schama, 1995; Gilroy, 2000). In both these cases, specific aesthetic genres of visual landscape are tied up with notions of self-improvement and self-transformation via travel and sightseeing. As McNaughten and Urry (1998) note, with the advent of touristic travel and the visual consumption of scenery through the nineteenth century, an ability to apprehend and comment upon landscape comes to signal taste, connoisseurship and cultural distinction. In this way, certain landscape forms and landscape experiences are made commensurate with certain elite forms of subjectivity. An especially influential credo here emerges from Romantic discourses of the self and nature, in which a commonly male subject undergoes rhapsodic or epiphanic experiences in the vicinity of a nature explicitly framed by the precepts of sublime aesthetics; a nature at once exotic, alluring, fearful, awesome and transformative (see Rose, 1993; Darby, 2000; Mills, 1996).

For Derek Gregory (1994), drawing on the work of Timothy Mitchell (1988) and, more generally, Martin Heidegger (1996), this interlocking of senses of self, forms of movement and styles of landscape visualisation produces the 'world-as-exhibition', a particular way of apprehending and representing the world as a visual spectacle. This, he argues, is how Western scientific and cultural discourse (including both the discipline of geography itself and diverse practices for visualising landscape) produces and judges knowledge, and understands both itself and others. Thus, 'it was a characteristic of European ways of knowing to render things as objects to be viewed and, as Timothy Mitchell puts it, to "set the world up as a picture . . . [and arrange] it before an audience"' (Gregory, 1994, p.34). In one sense this process can be literally witnessed in the creation of self-consciously spectacular imperial landscape stages such as the centre of London in the nineteenth century (Driver and Gilbert, 1998). Via pageants, tattoos and jubilees, the visual display of the British Empire's might, wealth and diversity upon a theatrical and monumental central London stage performed a world-straddling geography of self and other, centre and periphery, identity and difference. From the panoptical and authoritative vantage point of such imperial landscapes, the world symbolically unfurls itself. In this way, visual ordering and exhibiting of the world is inseparable from a colonial and imperial imagination in which an informed, rational and enlightened Western observer gazes upon an other simultaneously defined as exotic, irrational and so in need of ordering. In other work Gregory (1995, 1999) details how this gaze has produced the landscape of Egypt specifically as a visual exhibition to be consumed and appreciated; the general point, however, is that the confluence of voyaging and visualising can be understood as an imperial discourse 'precisely because it put in place an inside and an outside, a centre and a margin . . . this was a way of seeing the world as a differentiated, integrated, hierarchically ordered whole' (Gregory, 1994, p.36, original emphasis).

To conclude this section, however, it is worth noting most cultural geographers writing on landscape, travel and imperialism over the past fifteen or so years have been concerned to demonstrate how the holistic and totalising ambitions of the imperial landscape gaze were never, in fact, wholly realised. As was noted above, for Mitchell (1994a, p.10), the project of critically examining imperial landscape turned in part upon uncovering 'fractured images of unresolved ambivalence and unsuppressed resistance'. Equally, while many cultural geographers have sought to script landscape's imbrication within imperialist and colonialist projects – the landscape gaze *as* imperial discourse – so they have also sought to chronicle the slippages, betrayals and ambivalences integral to such projects. The critical history of imperial landscape thus aims to at once detail complicity *and* uncover absences and fissures within supposedly seamless canvases. In many cases this has involved the demonstration of ambivalence and complexity within the landscape gaze itself: ambivalence, for example, as regards the gendered status of the observer (e.g. Blunt, 1994) or in terms of conflicts between scientific and journalistic visions (Driver, 2001). But there has also been a growing recognition of the insufficiency of what Barnett (1998) calls a 'projection model' of colonial and imperial discourse; that is, a critical perspective which assumes that European visions of landscape were simply or straightforwardly projected onto or imposed upon non-European spaces. In actuality, as Pratt (1992) has noted, and as many others have sought to show, colonial and imperial landscape is precisely an encounter, a 'contact zone', an exchange and transformation; it is always already a hybrid landscape, no longer the preserve of either coloniser or colonised. Mitchell (1994a, p.27) himself finds this in nineteenth-century depictions of New Zealand, in which Maori culture is, for him, already 'tinting the *pakeha* [European] vision and decentring its imperial gaze. . . . The reading is of the encounter between two conventions, an encounter that leaves us in an odd, disturbing, liminal space, the threshold'.

## 4.6 CONCLUSION

This chapter has discussed some of the most significant substantive and theoretical directions explored by geographies of landscape through the 1990s and early 2000s. It has also been a lengthy chapter. Partly this is because the period in question has been such a fertile one for landscape studies in human geography. But partly also it is because this is a story that has not been brought together in this precise way before. One of my aims in this chapter has been to show how, through the 1990s, new and distinctive accounts of landscape were produced by cultural geographers and others – accounts distinctive, that is, from the understanding of landscape as a way of seeing discussed in Chapter 3. In this sense, the varied sets of writings discussed here can be understood as attempts to stake out quite different agendas for landscape.

Thus, Don Mitchell's work (Section 4.3) advances a more thorough-going, perhaps even more orthodox, materialist and Marxist vision of landscape than that found in the work of Denis Cosgrove, Stephen Daniels and James Duncan; one grounded in the material transformation of landscape through industrial and agricultural processes, and in the concomitant material transformation of human lives. Here, landscape is as much about production as consumption, labour as leisure; it works more through everyday economic and social processes than ideological tableaux. This notably politicised and radical account of landscape has little time for what it sees as the 'new cultural' removal of landscape to an aestheticised artistic and literary realm.

By contrast, work on 'cultures of landscape' (Section 4.4) advances a poststructural, Foucauldian cultural politics of power, subjectivity and conduct. While, in the work of David Matless and others, there is still a clear focus on issues of authority, visuality and claims to knowledge and ownership, landscape is conceived less as a cultural image or type of gaze and more as a migratory, 'shuttling' set of beliefs and practices, both formed by and informing senses of self. For Matless, the emphasis is on the discursive, that is on the symbolic and practical confluence of morality and subjectivity via landscape. Landscape is thus thought as a matter of conduct and forms of 'proper' bodily display and performance, as an ongoing, refracting set of claims and gestures:

> the question of what landscape 'is' or 'means' can always be subsumed in the question of how it works; as a vehicle of social and self identity, as a site for the claiming of a cultural authority, as a generator of profit, as a space for different kinds of living.
>
> (Matless, 1998, p.12)

Lastly this chapter has discussed some of the main themes of geographical writing on landscape, travel and imperialism (Section 4.5). Here, a major inspiration has been forms of postcolonial theory themselves indebted to Foucauldian and Derridean poststructuralism, with its characteristic stress upon presence, language, discourse and practices of division, expulsion and self-formation. In consequence, writing on landscape in imperial and colonial contexts has emphasised questions of representation, erasure and appropriation, the mapping-out of imaginative geographies of self and other, and issues of scientific and aesthetic authority in visual and textual landscape depiction. And throughout this literature, alongside a critical recognition of the complicity of discursive forms such as landscape with imperialist or colonialist systems, cultural geographers and historians have consistently sought to testify to tensions, complexities and unravellings; in doing so glimpsing a critical postcolonial politics.

In sum, a series of distinctive agendas for landscape opened up through the 1990s and this chapter has sought to capture these distinctions. Yet in concluding I would stress continuity as much as rupture. Particularly with regard to work on cultures of landscape and imperial landscape, the agenda of 'new cultural geography' – one linking landscape itself with visuality, ideology, authority, and landscape geographers with the critical episte-mologies and procedures of the arts and humanities – has been carried forward, extended and developed, and hopefully this is clear within the sections of this chapter. Not least because many key figures in these areas were the students and junior colleagues of writers such as Cosgrove and Daniels, what we see within geographies of landscape through the 1990s is a measure of evolution, and above all a continuing emphasis upon the critical interpretation of landscape in the form of visual and textual discourse. There is strong substantive continuity with respect to the foregrounding of notions of power, subjectivity, representation and visuality. It makes sense, therefore, to think of Chapters 3 and 4 of this book as a pair: together, they have sought to describe the main elements of the story of cultural geographies of landscape over the past twenty-odd years. At the end of these chapters the book now comes to something of a turning point. The next chapter considers what is in many ways a quite different understanding of landscape, one which takes inspiration from a branch of continental philosophy called phenomenology.

# 5

## LANDSCAPE PHENOMENOLOGY

### 5.1 INTRODUCTION

This chapter outlines and discusses work that uses the insights of phenomenology to write about landscape, and culture–nature relations more widely. In contrast to the forms of interpretative and discursive analysis discussed through the previous two chapters, landscape phenomenology often lays stress upon some measure of direct, bodily contact with, and experience of, landscape. It should be noted, however, that this does not simply mean a return to the more empirical 'field' approaches discussed in Chapter 2. Instead, as I will show, through adopting some distinctive theoretical perspectives, landscape phenomenology aims to offer something different from both empirical and discursive studies of landscape.

But I know from experience that for many people there is something unappetising about the very word 'phenomenology'. A difficult word to say – so by inference, it seems, it must refer to something itself difficult, complex and obscure. On the page, up on the screen, or on lips and tongue, phenomenology seems to begin with a stumble. People often assume or decide they will not be able to understand phenomenology. And implicit in this is a further assumption that phenomenology is something that can in fact be safely ignored or dismissed, because such a term must belong to a space of mere abstract theory, one far removed from everyday experience.

The irony is that a faithful description of everyday 'lived' experience is in fact one of the most long-standing and central goals of phenomenological writing. In the course of this chapter I want to make some elements of phenomenology clear. First, phenomenology is a significant branch of continental philosophy, first coming to prominence in the late nineteenth and early twentieth centuries through the work, in particular, of Edmund Husserl. Second, like most major philosophical traditions – for instance, like the various forms of materialism, structuralism and post-structuralism discussed in Chapters 3 and 4 – phenomenology has been defined and practised in diverse ways, and has developed and evolved over time, ramifying into a series of distinctive movements. One of these, most commonly called existential phenomenology, and closely associated with the work of Martin Heidegger and Maurice Merleau-Ponty, has been a principal source of insight and inspiration for a raft of recent writing on landscape, culture–nature relations, embodiment and subjectivity.

This chapter discusses the recent re-emergence of phenomenological perspectives within both cultural geographies of landscape and cognate disciplines such as cultural anthropology and interpretative archaeology. I say 're-emergence' because human geography has already experienced an engagement with phenomenology, via the humanistic geographies of the 1970s. This will be discussed again later; to begin with, however, in order to first clarify the core precepts and suppositions of landscape phenomenology, the chapter will outline some of the most significant arguments advanced by writers in this area, in particular notions of being-in-the-world and embodiment (Section 5.2). (Box 5.1 supplies another introduction to phenomenology, through a study of a particular work of art.) Forms of phenomenological understanding in turn enable a critique of the epistemologies underlying both the definition of landscape as a way of seeing, and the interpretative or constructivist paradigm informing much cultural geography. This critique has been presented most sharply by the cultural anthropologist Tim Ingold (2000), a key advocate of phenomenological approaches to landscape, and Section 5.3 turns to discuss his work in some detail. This provides the springboard from which the next section of the chapter (5.4) discusses how, as part of a more general re-focusing upon notions of corporeality, practice and performance, cultural geographers and others have in recent years used phenomenological modes of understanding to address issues of landscape, nature, embodiment and performance. The final substantive section of the chapter, however, outlines

## Box 5.1 *SPIRAL JETTY*

Phenomenological approaches often stress direct, bodily contact with, and experience of, landscape. They commonly aim to reveal how senses of self and landscape are together made and communicated, in and through lived experience. This box discusses the making of self and landscape through the example of the creation of a noted piece of Land Art.

Land Art, or Earth Art, is a broad artistic movement which began in North America in the late 1960s. As its name implies, Land Art involves the use and manipulation of physical materials and environments as a means of creative expression. UK-based land artists such as Andy Goldsworthy and Richard Long, for example, utilise found objects such as rocks and leaves, and natural processes of growth, change and decay, in the creation of art works which are conceived, executed and situated in outdoor and remote locations such as moors, hillsides and beaches.

As Kastner and Wallis (1998) relate, one of the initial aims of the Land Art movement was to liberate landscape art from galleries and museums, and from confined and controlled settings in general, and so to take artistic practice outdoors – into 'natural' or relatively untouched spaces in one sense, but also into marginal or neglected areas such as freeways, industrial riversides and despoiled and polluted sites. This movement outdoors signalled both a conscious rejection of the commercialism of the mainstream art world and a dawning awareness of environmental stresses and vulnerabilities. And these beliefs and values further chimed with the emerging radical world views of sixties counter-culture.

In terms of practice and performance, we can argue that much Land Art involves a physical, tactile and dynamic relation between artist and site, one in which landscape is sensed and represented as a creative and elemental force in its own right. Robert Smithson, one of the leading figures of the Land Art movement in its early years, explores precisely this active notion in perhaps his best known work, *Spiral Jetty* (viewable at: http://www.robertsmithson.com/earthworks/spiral_jetty.htm)

continued

Smithson's *Spiral Jetty*, constructed in 1970, juts out into the Great Salt Lake in Utah, in the north-west United States. It is monumental in scale, a vast shape set in the middle of a harsh, barren, rocky geology. It is composed of 6,783 tonnes of rock, earth and salt crystals, all of which took 625 hours to move into position, with a succession of trucks and tractors creating the artwork by progressively dumping their loads out into the lake along the outline of the spiral staked by Smithson (Kastner and Wallis, 1998).

As a work of art, the *Spiral Jetty* may be approached in two ways. First you can view it as a whole, from an elevated viewpoint. Smithson himself filmed a video with commentary from a helicopter circling above the completed jetty. Second, however, you can approach it from ground level, and immerse yourself in the work and its surroundings by walking out along the jetty. In other words the *Spiral Jetty* is both something to look *at* and also a platform *from* which to experience the surrounding environment. It is a landscape that creates and enables a number of different relationships and connections between observer, observed and environment. You can look at it, look from it, look with it, be in it, be part of it, connect it up with yourself and the surroundings in a number of different ways.

Smithson wrote that a sense of being immersed within and overwhelmed by the intense colours and shapes of the locality inspired *Spiral Jetty*:

> As I looked at the site, it reverberated out to the horizons only to suggest an immobile cyclone, while flickering light made the entire landscape appear to quake. A dormant earthquake spread into the fluttering stillness, into a spinning sensation without movement. This site was a rotary that enclosed itself in an immense roundness. From that gyrating space emerged the possibility of the *Spiral Jetty*. No ideas, no concepts, no systems, no structures, no abstractions could hold themselves together in the actuality of that evidence. . . . It was as if the mainland oscillated with waves and pulsations, and the lake

remained rock still. The shore of the lake became the edge of the sun.

(cited in Kastner and Wallis, 1998, p.215)

What Smithson is suggesting here is that the *Spiral Jetty* was as much a product of the land as of his own imagination. The form of the spiral, a sense of gyrating movement, was already immanent in the landscape, an energy waiting to be released; the spiral was not just an arbitrary form imposed upon a blank canvas by the artist. To put this another way, the *Spiral Jetty* as landscape is the material, tangible expression of a certain fusion or connection between mind and matter, an intense, spiralling relationship between Smithson and the land.

In *Spiral Jetty*, the emphasis is thus upon two aspects of landscaping relations. First of all we gain a sense that nature, or the land itself, in terms of topography, atmosphere, geology, is a powerful, elemental and active agent, co-constructing the artwork. The land, Smithson suggests, is not a *tabula rasa* waiting to be etched with whatever dreams or visions we might have: it shapes us as we shape it. *Spiral Jetty* is deliberately open to change and decay, for example the rising level of the Great Salt Lake into which it juts means that today it only periodically appears above the surface of the waters.

Second, the emphasis of *Spiral Jetty* is upon immersion in and corporeal experience of landscape. Instead of being a static scene to survey with a cool, measured and discerning gaze, landscape here is mobile and multi-sensory; it surrounds us as well as being in front of us. Both Smithson's words and the material form of *Spiral Jetty* take us in the direction of what are called phenomenological approaches to, and understandings of, landscape. In these approaches landscape is conceptualised in terms of active, embodied and dynamic relations between people and land, between culture and nature more generally. The general argument is that landscape comprises the totality of relations between people and land. These relations are seen as ongoing and evolving rather

continued

> than static, they constitute an embedded and engaged being-in-
> the-world that comes before any thought of the world or of
> landscape as merely an external object. Body and environment fold
> into and co-construct each other through a series of practices and
> relations. The *Spiral Jetty*, with its emphasis upon the viewer's
> multi-sensory immersion within the work, and Smithson's account
> of an almost-primordial connection with the earth, give an initial
> clue as to how such an understanding of being-in-landscape might
> be exemplified.

some of the problems and difficulties arguably inherent in a purely or
classically phenomenological approach.

## 5.2 INTRODUCING PHENOMENOLOGY: FROM DISEMBODIED GAZE TO LIVED BODY

### 5.2.1 René Descartes: vision and knowledge

One means of introducing the main tenets and arguments of phenomenology is to think back to the notion of landscape as a *way of seeing the world*
– discussed at length through the previous two chapters. To rehearse again
ideas which should by now be familiar, once defined as a way of seeing,
landscape is situated within the perspectival traditions of Western art,
and so becomes cognate with the figure of a gazing spectator (Berger, 1972;
Cosgrove, 1985). Perspectival techniques of landscape depiction allowed
artistic representation to approximate the realities of perceptual experience,
by offering on the canvas a 'realistic' and plausible portrayal of the visible
world. In this way, the idea of landscape is linked to notions of visual
observation, detachment and objective knowledge. Landscape thereby
becomes associated with an understanding of knowledge as something
both *produced by* and *located in* a detached, observing subject, a subject who,
through and by the depth of perspective, is able to stand aloof from
the dramas and intricacies of a objective world positioned 'beyond' and
'outside'. To put this another way, as a way of seeing, landscape becomes
the accomplice and expression of what is often called a spectatorial

epistemology, an approach to knowledge which begins by supposing the following scenario: an external pre-given reality observed and represented from a detached position by an independent perceiving human subject.

These statements may seem abstract, however it can easily be shown that connections between vision and knowledge in particular are deeply implicated in the everyday or 'commonsensical' ways in which we define both ourselves and our relation to the external world. Seeing is colloquially associated with accurate, reliable and objective knowledge. In this vein, Martin Jay (1993) opens *Downcast Eyes*, his comprehensive account of the philosophical history of vision, by offering a demonstration of just how deeply embedded visual metaphors are in everyday language. When we express an opinion, we say that's my point of view. If we wish to know what another thinks about a given topic, we ask for their perspective on it. If we agree with them, we say I see what you mean. In order to attain reasonable, rational and measured knowledge, we need to stand back and get things in perspective. In sum, we define ourselves – we define the essence of what it is to exist as a human being – in terms of visual detachment. We define ourselves not as creatures in a world but as points of view upon it, as spectators looking at it from a distance, or from above.

It is precisely this understanding of subjectivity and the external world that phenomenology opposes and aims to supplant. The founder of modern phenomenology, Edmund Husserl, called such an understanding the 'natural attitude', by which phrase he meant, in part, our ingrained and taken-for-granted habits of considering ourselves, first, as discrete subjects with internally realised thoughts and feelings, and the world, on the other hand, as an external and real object. For Husserl, the natural attitude underwrote many of the procedures of Western scientific practice, in particular its assumption of an external objective reality amenable to accurate measurement and description. Equally we can pinpoint the invention, via perspective, of the landscape way of seeing as another pivotal moment in the evolution of the modern Western sense of self. But the most commonly identified source of this spectatorial epistemology is the philosophy of René Descartes, which emerged almost two hundred years after the advent of perspective.

To abbreviate substantially, Descartes's system of methodical doubt, as outlined in his *Meditations* (1975 [1641]), is an attempt to banish the possibility of error in the human perception and representation of the visible world. The process begins in the 1st Meditation, where the quest for certain knowledge leads Descartes to posit 'an evil spirit, who is supremely powerful and intelligent, and does his utmost to deceive me' (1975, p.65). To be certain, therefore, he must 'suppose that sky, air, earth, colours, shapes, sounds and all external objects are mere delusive dreams' (ibid.), and that thought provides the only solid anchorage for knowledge. Thus Descartes argues that the senses are inherently deceptive. We cannot always trust the evidence of our eyes, ears, fingers and so on. We might get things wrong. We might be deceived. Or perhaps we are hallucinating, or dreaming. The senses are uncertain, ambiguous in the messages they send us, and so cannot provide a basis for certainty – for knowledge that is, in Descartes's terms, 'clear and distinct'.

For Descartes, the one and only thing we can be indubitably sure of is that we are thinking beings. For, even when we think our senses are deceiving us, we are still thinking. Thinking is thus the essence of being human, or, as Descartes puts it, 'I think therefore I am'. Cogito ergo sum. The price of certainty is thus an absolute distinction between mind and body, and thought and world. The thinking, rather than sensing, subject becomes the sole arbiter of certainty. Crucially, however, Descartes's philosophical language and method clearly identifies vision with this cogito, and thereby cements deep associations between seeing and thinking, and visual perception and certainty. From these formulae there devolves a series of significant binaries and dualisms, for instance:

Vision/Touch
Thought/Senses
Mind/Body
Subject/Object
Inside/Outside
Culture/Nature
res cogitans/res extensa.

With these dualities Descartes concludes a journey that begins with perspective's separation of observer and observed. The implications of the landscape way of seeing for the gazing subject are formalised and clarified. Vision becomes definitively associated with reason and the mind. The gazing subject becomes the locus of agency and reason. The external landscape in turn is rendered as inert matter, whose sole quality is extension in three dimensions. Having argued that sensory perception is inherently 'deceitful', the means whereby vision alone can testify to the geometrical certainties of the material world thus necessarily involve its removal to an idealised, uncontaminated vantage point. A certain *distance* is established, carrying the gazing subject from the world to a position of detached epistemological authority, and then allowing that gaze to contemplate the world along an avenue of inspection and observation.

### 5.2.2  Maurice Merleau-Ponty: perception and the lived body

It is worth pausing here for a moment, and reiterating just how influential Cartesian philosophies have been with regard to Western systems of thought, science and visual cultures. Dualities such as mind/body, culture/ nature, thought/reality are the very categories and distinctions we live within, think with, assume as our basis for judging what is real and true. And so in seeking to supplant these, phenomenology grapples with fundamental issues around subjectivity, knowledge and perception. It seeks to show above all that the Cartesian perspective does *not* in fact truthfully describe lived, human experience – the human experience of landscape, for instance.

A key source of inspiration for cultural geographers and others using phenomenological notions to talk about the human experience of being in landscape has been Martin Heidegger's (1996) work on dwelling and 'being-in-the-world'. Section 5.3 will discuss this in relation to Tim Ingold's (2000) work. Here, however, I want to foreground the work of another writer, Maurice Merleau-Ponty (1962, 1968), as a way of introducing some of the central themes of phenomenology with especial reference to questions of vision, knowledge and embodiment – questions central to landscape phenomenologies.

The significance of Merleau-Ponty's work for landscape studies in cultural geography and beyond lies in the manner in which he returns, time and again, to the entwined topics of *vision* and *embodiment*. Merleau-Ponty is, above all else, *the* philosopher of the body, and his writing can be understood first of all as a sustained attempt to overturn persistent Cartesian biases in Western philosophy in favour of the mind, reflection and cognitive representation, by instead insistently attending to and describing the indelibly corporeal nature of human being, knowledge, experience and perception. In this regard, his work may be usefully, if a little artificially, divided into two phases, corresponding to two major works, the *Phenomenology of Perception* (1962 [1942]) and the unfinished *The Visible and the Invisible* (1968 [1961]).

Merleau-Ponty's project is quintessentially anti-Cartesian. He argues that the body is the basis and conduit of knowledge, rather than something that has to be disregarded in order to attain 'true' or 'certain' knowledge. For Merleau-Ponty, our lived bodies *cannot* be disregarded. To do so would be to 'forget' the world of concrete existence, and enter instead a secondary realm of theoretical reflection and abstraction. The lived body is therefore the beginning and the end of Merleau-Ponty's philosophy – it is both that which is to be described *and* that which any description presupposes. Human being is being embodied. Thus the *Phenomenology of Perception* declares that the lived body, my body, is 'always there for me' (Merleau-Ponty, 1962, p.90). And this involves much more than an acknowledgement of the fact of embodiment. Here, the body is never simply an instrument or object which 'I' have permanently and conveniently at my disposal. As Elizabeth Grosz (1994, p.86) precisely notes, for Merleau-Ponty, 'the body is not an object. It is the condition and context through which I am able to have a relation to objects'. Thus, what his work describes is a *body-subject*, an always-already-incarnate subjectivity, a self inseparable from its embodiment. There can be no notion of an ephemeral 'mind' or 'self' which somehow inhabits the body, and thus constantly 'experiences' it. As Nick Crossley (1995, p.44, emphasis in original) clarifies: 'It is important to stress . . . that Merleau-Ponty's account should not be read as an account of our experience *of* embodiment. Embodiment is not experienced in this account. It is the very basis of experience.'

In one sense, therefore, Merleau-Ponty's aim is to replace the 'objective' Cartesian body, conceived empirically as mere mechanical substance, with the body-as-subject, and so to reinvigorate the bodily as the locus and precondition of subjective action. From the start, my body is the very basis of my intention and awareness; it is not a puppet figure animated by directives and representations emanating from a disembodied consciousness. To put this another way, 'bodily space is not space thought of or represented' (Merleau-Ponty, 1962, p.137). That is, I do not in the first place have an 'idea' of my body which I then put to work. Rather 'experience of one's own body runs counter to the reflective procedure which detaches subject and object from each other and which gives us only thought about the body, or the body as an idea' (*ibid.*, p.190).

In foregrounding engaged experience, rather than detached reflection, Merleau-Ponty's conception of being-embodied may be used as a way of understanding the classical phenomenological notion of being-in-the-world. I am my body, which is always already both in and of the world. The body's active agency within the world thus does not consist of a series of operations upon a pre-given space. Rather the body is both always already immersed in worldly spatiality, and also creative of that space; it is, as Edward Casey (1998, p.229) puts it, 'spatiogenetic'. There is forever performed 'a certain possession of the world by my body, a certain gearing of my body to the world' (Merleau-Ponty, 1962, p.250). Unequivocally, then, 'far from my body being for me no more than a fragment of space, there would be no space at all for me if I had no body' (ibid., p.102). The body's spatiality constitutes an anchorage within and by the world: the Cartesian disembodied gaze is, for us, a literal impossibility. 'My body is a thing amongst things, it is caught in the fabric of the world' (Merleau-Ponty, 1969, p.256).

### 5.2.3 Landscape, vision and embodiment

Pausing once more at this point, it is worth noting that Merleau-Ponty's initial phenomenology of self, body and world already enables a new definition of landscape as a term and concept. Divested of assumptions regarding observation, distance and spectatorship, the term landscape ceases to define a way of seeing, an epistemological standpoint, and instead becomes potentially expressive of being-in-the-world itself: landscape as a milieu of engagement and involvement. Landscape as 'lifeworld', as a world to live in, not a scene to view.

This, in large measure, was the understanding of Merleau-Ponty's philosophy, and of phenomenology in general, taken up by the humanistic geographies of the 1970s. As Chapter 2 noted, the humanist ethos, as exemplified by J.B. Jackson's (1976, p.2) argument that 'far from being spectators of the world we are participants in it', was very much tied up with a vision of landscape as a shared, lived-in world. This sensibility can be seen carried forward in collections edited by Meinig (1979b) and more recently Thompson (1995). Another especially notable example of humanistic geography's phenomenological bent is David Seamon's (1979) well-known work on 'body ballets' and 'place ballets', which sought to describe the everyday worlds of individuals in terms of embodied phases of movement, rest and encounter. As Pickles (1985) notes, however, a difficulty with this and other such pioneer accounts in human geography (e.g. Buttimer, 1976) is their tendency to equate phenomenology with the description of subjective, lived experience – and with a valorisation of 'subjective meaning' in general. This assumption of a correlation between phenomenology and a supposedly ahistorical and discrete subjectivity is still sometimes made by more contemporary writers (e.g. Nash, 2000; Cresswell, 2003). As the preceding pages should have shown, however, phenomenology is better conceived as a radical re-visioning of our received (i.e. Cartesian) notions of subjectivity and subject–object relations. Moreover, Merleau-Ponty's analysis does not simply relocate the self in the body, and then the body 'in' the landscape. It goes beyond a redescription of landscape as bodily lifeworld, and becomes notably sharper in so far as it explicitly seeks to redefine *vision* in corporeal terms.

Again Descartes's account of the gazing/thinking subject forms the target. Merleau-Ponty first highlights some of the dilemmas and contradictions introduced by this account, in noting that 'Descartes tells us that the existence of visible things is doubtful, but that our vision, when considered as a mere thought of seeing, is not in doubt' (1962, p.375). Elsewhere, he describes the Cartesian account of vision as the 'mind's eye' as 'the breviary of a thought that wants no longer to abide in the visible world and so decides to construct the visible according to a model in thought' (Merleau-Ponty, 1969, p.263). For Merleau-Ponty, by contrast, there can be no question of *not* 'abiding in the visible'. Human being, as being-embodied, is forever anchored within the visible world, through that embodiment. The embodied vision of the individual subject is thus precisely a particular point of view *within* the world – not a gaze from without.

Human vision is always 'caught in the fabric of the world'. That is, vision is not the particular hallmark of a detached, spectating subject. From the start, Merleau-Ponty argues, we are engaged *with and by* the world around us. Thus, in common with the other senses, our vision in no way emerges from a detached vantage point. When we look, what is occuring is an *enlacing together* of body and world:

> As I contemplate the blue of the sky I am not set over against it as an acosmic subject; I do not possess it in thought, or spread out toward it some idea of blue . . . I abandon myself to it and plunge into this mystery, it thinks itself in me.
>
> (Merleau-Ponty, 1969, p.214)

In Merleau-Ponty's final, unfinished text, *The Visible and the Invisible* (1968 [1961]), these phenomenological accounts of subjectivity, vision and embodiment are taken further forward and, in some ways, developed quite radically. Here, he develops what is called the thesis of reversibility. This is a complex-sounding but in fact decidely grounded argument. Reversibility, for Merleau-Ponty, refers to the fact that the body is always *both* subject and object. We can always as subjects perceive parts of our bodies as objects, as revealed through the example of one hand touching another. Here, one hand is the subject (or toucher), while the other is the object (that which will be touched). And yet, as experience shows, at the moment of the touch, these roles become indistinguishable from each other – or *reversible*. Our bodies can be – in truth, they *are* – both touching and touched, observer and observed, active and passive, subject and object.

Merleau-Ponty argues this notion of the body's reversibility is evident as much in the visual as the tactile. For I both see and am seen. And this reversibility of vision fundamentally alters the relationship between the 'perceiving subject' and the 'perceived world'. It is, Merleau-Ponty (1968, pp. 130–131) writes, 'as though our vision were formed in the heart of the visible, or as though there were between it and us an intimacy as close as between the sea and the strand'. This intimacy involves much more than simply saying the subject's vision is *close to* the visible world of things. For Merleau-Ponty 'he [sic] who sees cannot possess the visible unless he is possessed by it, unless he *is of it* . . . he is one of the visibles, capable, by a singular reversal, of seeing them' (ibid., pp. 134–135, emphasis in original). To put this another way, the self is not simply 'in' the world – it is *of* it:

> As many painters have said, I feel myself looked at by the things, my activity is equally passivity – not to see in the outside, as the others see it, the contours of a body one inhabits, but especially to be seen by the outside, to exist within it, to emigrate into it, to be seduced, captivated, alienated by the phantom, so that the seer and the visible reciprocate one another and we no longer know which sees and which is seen.
>
> (Merleau-Ponty, p.139)

Merleau-Ponty uses the term *intertwining* to capture the way in which self and landscape relate to each other. As my body is both observer and observed, seer and seen, its relations with the visible world intertwine in a double movement of separating and joining. Joining because I *can be seen*, I am part of the visible landscape. Separation because I *see*; as Michel De Certeau (1983, p.26) remarks 'we cannot open our eyes to things without distancing ourselves from what we seek. Separation is the price of vision'.

In this way, Merleau-Ponty offers an original conception of embodied vision as an enlacement or intertwining of self and landscape. This releases the visual gaze from its detention as the accomplice of Cartesian spectatorial epistemology. The visible landscape, for Merleau-Ponty, is neither the 'field of vision' of an observing subject, nor simply the sum total of external visible things. The visible landscape is instead an ongoing process of inter-twining *from which* my sense of myself as an observing subject emerges. It is the fact that I belong to the landscape of visible things that enables my seeing – it is my seeing which enables me to witness that belongingness. And so subjectivity, and the possibility of meaningful engagement with the visible world, occurs as the arising of a 'point of view' within the visible – it is thus *produced* ever and anon within embodied practices. As Merleau-Ponty (1968, pp.137–138) concludes, the subject 'is neither thing seen only nor seer only, it is visibility sometimes wandering and sometimes reassembled' (please note that my own work on landscape, subjectivity and the visual (e.g. Wylie, 2002a, 2005, 2006) has tried to explore some of these Merleau-Pontyian ideas in the context of specific locales and bodily practices).

To conclude this section, it is worth reiterating the argument that Merleau-Ponty's philosophy is especially salient as regards concepts of visual landscape and the gaze because, as Martin Jay (1993) details, it remains the most sustained and ambitious philosophical attempt to provide an alternative to Cartesian accounts of vision. For Merleau-Ponty, landscape is not a way of seeing: the terms refers instead to the materialities and sensibilities *with which* we see (Wylie, 2005, 2006). When I look, I *see with* landscape. I am neither looking *at* it, nor straightforwardly placed 'inside' it. I am intertwined instead within an unfolding differentiation. To put this perceptually, I perceive through an attunement with landscape (see Lingis, 1998).

Merleau-Ponty's work thus transforms notions of landscape, vision and subjectivity. Landscape moves from a particular type of knowing (a way of seeing), to a specific mode of being (a seeing-with) – from a Cartesian spectatorial epistemology to a phenomenological ontology. The gazing subject is no longer, as Descartes claimed, an a-priori 'thinking' self projecting meaning onto the landscape. Instead it is a self assembled and performed via bodily practices of landscape:

> The painter's vision is not a view upon the *outside*, a merely physical-optical relation with the world. The world no longer stands before him [*sic*] through representation: rather it is the painter to whom the things of the world give birth by a sort of concentration or coming-to-itself of the visible.
>
> (Merleau-Ponty, 1969, p.276, emphasis in original)

## 5.3 LANDSCAPE AND DWELLING

### 5.3.1 The critique of landscape as a 'way of seeing'

This section will discuss how the notions of phenomenology, landscape, embodiment and vision outlined thus far may be taken forward to both critique extant notions of landscape and formulate a fairly systematic alternative, one based based on an equation of landscape with human dwelling-in-the-world. In 'The temporality of the landscape' (Ingold, 1993, reprinted in Ingold, 2000), a paper which has become something of a cardinal citation, the cultural anthropologist Tim Ingold puts forward a series of criticisms of notions of landscape as a way of seeing the world – landscape as an image, representation or gaze composed of specific cultural values and meanings. He also offers an alternative understanding, based upon insights from phenomenology as well as from ecological psychology and from anthropology itself. Here, then, I will examine Ingold's arguments in some detail. It should be noted first of all that these are part of a much wider and longer-term project to challenge received anthropological understandings of perception and culture–nature relations; and so in order to more fully contextualise this understanding of landscape I want to draw upon material from some of the other essays collected together in *The Perception of the Environment* (Ingold, 2000) – the volume in which 'The temporality of the landscape' is reprinted.

As Chapter 3 discussed, new cultural geographies were concerned to develop analyses that implicated landscape images and texts within systems of cultural, political and economic power. In doing so, they defined landscape as a particular set of cultural values, attitudes and meanings: a 'way of seeing' the world. Yet it can be argued – and this is what Tim Ingold *does* argue – that while such a definition might be critically helpful, on a more fundamental level it divisively enshrines and perpetuates a series of dualities – between subject and object, mind and body and, especially, between culture and nature.

For example, in their introduction to the influential collection, *The Iconography of Landscape*, Denis Cosgrove and Stephen Daniels (1988a, p.1) define landscape as 'a cultural image, a pictorial way of representing or symbolising surroundings'. But, citing this definition, Ingold (2000, p.191) declares:

> I do not share this view. To the contrary, I reject the division between inner and outer words – respectively of mind and matter, meaning and substance – upon which such distinction rests. The landscape, I hold, is not a picture in the imagination, surveyed by the mind's eye, nor however is it an alien and formless substrate awaiting the imposition of human order.

Therefore for Ingold the definition of landscape as a 'cultural image', that is as first and foremost a symbolic representation, is profoundly schismatic. This is because it involves positing on the one hand a set of disembodied cultural meanings – a symbolic landscape – and on the other a bare, blank bedrock – a physical landscape – onto which cultural meaning is projected. In other words the definition of landscape as a cultural image, or as a 'way of seeing', assumes and reproduces a fundamental distinction between the *ideas of culture* and the *matter of nature*. And Ingold makes the further point that this sort of distinction informs most academic research and writing not only in cultural geography, but throughout social and cultural theory, and especially throughout social and cultural anthropology, his own primary domain. His citation of Cosgrove and Daniel's work thus locates such geographies with a wider interdisciplinary context of landscape and culture/nature studies, one embracing archaeology, biological and cultural anthropology and (eventually) Western philosophy as a whole. Within and across these disciplines he detects the persistent and, for him, pernicious influence of dualistic thinking. As an example we could point to the commonplace assumption that the world may be divided into 'natural' and 'cultural' (or human) elements. But above all else Cartesian dualism may be witnessed in the academic division of labour between the natural and social sciences. This division is a particularly sharp issue for geography (physical and human) and anthropology (biological and social or cultural). As a prime consequence of this division, Ingold writes, the conceptualisation of landscape has persistently been bedeviled by a, 'sterile opposition between the naturalistic view of the landscape as a neutral, external backdrop to human activities, and the culturalistic view that every landscape is a particular cognitive or symbolic ordering of space' (ibid., p.189).

Throughout *The Perception of the Environment*, Ingold is less concerned with discussing scientific or 'naturalistic' understandings of landscape than with contesting the 'culturalist' epistemologies prevalent across much of anglophone humanities and social science research over the past twenty to thirty years. The term 'culturalist' here covers a range of movements and interpretative methodologies; for example semiotics, hermeneutics, iconography and some forms of discourse or textual analysis. It is worth recalling that, as discussed in Chapter 3, such movements clearly underwrote and guided the initial phases of the 'cultural turn' in human geography (e.g. see Cosgrove and Daniels's (1988) advocacy of iconography, Duncan and Duncan's (1988) use of Barthean semiotics, Duncan and Ley's (1993) use of hermeneutics). For Ingold, from an anthropological perspective, key examples of these sorts of approaches would include Clifford Geertz's (1973) *The Interpretation of Cultures*, and Clifford and Marcus's (1986) highly influential *Writing Culture: The Poetics and Politics of Ethnography*.

Linking together all of these various works is, first, an understanding of culture as an in-essence immaterial realm of texts, images, signs, symbols, representations, discourses and so on. Second, the 'construction' of cultural meaning is conceived as a movement of inscription or signification through which the material and concrete planes of the world (e.g. the body, nature) are given life, value and resonance. Reality is, so they say, culturally constructed and communicated via text, image and representation; via a swirl of accumulated, shared codes, signs and meanings.

Ingold's (2000) collective term for this understanding is the 'building perspective'. This may be defined most straightforwardly as the supposition that 'worlds are made before they are lived in' (ibid., p.179). More expansively, Ingold defines the building perspective as revolving around 'the premise that human beings inhabit discursive worlds of culturally constructed significance, laid out upon the substrate of a continuous and undifferentiated physical terrain' (ibid., p.172). In other words, this perspective assumes and reproduces a duality of cultural mind and physical nature. And in doing so it also draws a line between human and natural worlds. As Ingold further writes, 'to suggest that human beings inhabit discursive worlds of culturally constructed significance is to imply that they have already taken a step out of the world of nature' (ibid., p.14).

The often-implicit presence of the building perspective is, for Ingold, at the heart of a series of intellectual dilemmas facing academic disciplines such as human geography, anthropology and archaeology. For a start, there is the problem of thinking itself: 'the starting-point, in all such accounts, is an imagined separation between the perceiver and the world, such that the perceiver has to reconstruct the world, in the mind, prior to any meaningful engagement with it' (ibid., p.178). Form, significance and intentionality are thereby assumed to be confined to an anteriorised and interiorised space of thought, and more broadly culture. As Merleau-Ponty's work showed, this poses a problem as regards understandings of the body in particular. It becomes a puppet figure: the building perspective leads inexorably towards 'a tendency to treat bodily praxis as a mere vehicle for the outward expression of meanings emanating from a higher source in culture' (ibid., p.169). And because bodily praxis, action and performance are assumed to be secondary effects, rather than primary milieu, all accounts of cultural meaning emanating from the building perspective tend to prioritise the activities of a detached consciousness, and therefore presume a basic distinction between mental and physical worlds, between mind and body, thought and matter. Thus, Ingold argues, for Western epistemologies in general, 'meaning does not lie in the relational contexts of the perceiver's involvement in the world, but is rather laid over the world by the mind' (ibid., p.51).

In the development of Ingold's argument, the consequences of this building perspective epistemology are seen as having been especially problematic for the Western discipline of anthropology, particularly in its focus upon the study of non-Western cultures. The primary difficulty is that Western anthropologists have tended, by and large, to reproduce Western dualities of mind/body and culture/nature in their accounts of non-Western indigenous practice. For example, Ingold argues, indigenous accounts may sometimes work through a set of beliefs wherein order and meaning are understood to be immanent within landscape and nature themselves – as is the case in some Australian Aboriginal cultures. However, when Western academics come to interpret indigenous narratives, 'we find a complete inversion, such that meanings that people claim to discover in the landscape are attributed to the minds of the people themselves, and are said to be mapped *onto* the landscape. And the latter, drained of all significance as a prelude to its cultural construction, is reduced to *space*, a vacuum to the plenum of culture' (*ibid.*, p.54, original emphasis).

Thus, one particular consequence of a culturalist building perspective is the installation of a hierarchy which reductively positions the material aspect of landscape as an inert physicality, a content bereft of form. Turning back towards landscape geographies specifically, the irony here is that new cultural geographies of landscape were guilty of precisely the same offence that they tended to accuse the landscape concept, as a particular Western historical and cultural formation, of committing. Their critique suggests that landscape, as a 'way of seeing', is cognate with a Cartesian spectatorial epistemology which severs subject from object, mind from matter, culture from nature, and that this is evident in the elitist, proprietorial and imperialist lineaments of Western landscape discourse. The problem is that once landscape is defined as a 'cultural image', as a set of cultural meanings, and is understood as being formed first within a disembodied realm of thoughtful cultural discourse and only then projected outward onto the bare matter of the world, then we are, once again, firmly in the grip of Cartesian dualism.

## 5.3.2 The dwelling perspective

Ingold's solution to these various dilemmas involves a turn towards phenomenology, and the elaboration of what he terms 'the dwelling perspective'. As was briefly mentioned in the previous section, this notion of dwelling comes primarily from the later work of Heidegger (1996), and in particular his essay 'Building dwelling thinking', in which dwelling is positioned as an alternative, more holistic means of expressing human being-in-the-world. Colloquially, the word dwelling suggests ideas of home and inhabitation, and Heidegger's understanding is also rooted in these understandings. In this way, as Cloke and Jones (2001, p.651) describe it,

> Dwelling is about the rich intimate ongoing togetherness of beings and things which make up landscapes and places, and which bind together nature and culture over time. It thus offers conceptual characteristics which blur the nature/culture divide, [and] emphasise the temporal nature of landscape. ·

More fundamentally, we might say, for Heidegger dwelling describes in a total sense 'the manner in which mortals are on the earth' (Heidegger, 1996, p.350). No commas insert themselves between building and dwelling, or between dwelling and thinking; these three terms have become, for Heidegger, equal and synonymous. Thus, as Paul Harrison (2007, forthcoming) notes, the term dwelling encapsulates the attempt of an entire phenomenology to transcend a subject–object model of life and thought:

> the thought of dwelling begins as little with the world, the outside and the objective as it does with the ego, individual or the subjective. . . . Dwelling denotes [instead] the inflection from out of which these horizons emerge, it names a folding and unfolding of space, a twisting and criss-crossing of interiority and exteriority from which both these directions gain their sense.

Ingold the anthropologist chooses to describe dwelling first and foremost in terms of *practical activity*, by both humans and animals, that is rooted in an essential, ontological engagement with the material environment. 'It is through being inhabited', he writes (2000, p.173) writes, 'that the world becomes a meaningful environment.' The dwelling perspective thus focuses upon 'the agent-in-its-environment, or what phenomenology calls "being-in-the-world", as opposed to the self-contained individual confronting a world "out there"' (ibid.). And in focusing upon the 'agent-in-its-environment', upon ongoing, relational contexts of involvement, the dwelling perspective seeks to deny and dispel the tenets of dualistic thought – its separation of culture from nature, the discursive from the material. In particular, Ingold notes, dwelling re-visions the human being as,

> a being immersed from the start, like other creatures, in an active, practical and perceptual engagement with constituents of the dwelt-in world. This ontology of dwelling, I contend, provides us with a better way of coming to grips with the nature of human existence than does the alternative, Western ontology whose point of departure is that of a mind detached from the world.
>
> (Ingold, 2000, p.42)

Drawing further upon the work of the anthropologist Gregory Bateson, Ingold argues that dwelling necessitates a quite different account of mind

– of how thought operates and knowledge is produced and communicated. From the dwelling perspective, 'mind should be seen as immanent in the whole system of organism–environment relations in which we as humans are necessarily enmeshed, rather than as confined within our individual bodies *as against* the world of nature "out there"' (*ibid.*, p.16, original emphasis). This 'ecological' (Gibson, 1979) account of thought and perception thus runs counter to the common concept of thought as a passive, reflective 'armchair' activity, one that takes place on the inside, in the mind, indoors and so on. Instead, thought and knowledge become active and engaged; they occur through interactions between people, and interactions between people and environments. Importantly, neither 'people' nor 'environments' are constructed as fixed, stable, already-given entities here. Both are rather seen as continually developing and elaborating via interactions. In other words, the *activities* of dwelling are primary and incessant. Thus, 'in this approach, both cultural knowledge and bodily substance are seen to undergo continuous generation in the context of an ongoing engagement with the land and with the beings – human and non-human – that dwell therein' (Ingold, 2000, p.133).

Equally, from the dwelling perspective, the terms 'nature' and 'environment' no longer refer to a mute, physical world external to human thought. In particular, nature ceases to be seen as inert matter or 'bedrock' to be inscribed with meaning. Instead, 'far from being inscribed upon the bedrock of physical reality, meaning is immanent in the relational contexts of people's practical engagement with their lived-in environments' (*ibid.*, p.168). And such an account of culture–nature relations, it should be noted, is not about simply re-describing culture as practice and engagement, rather than representation and signification, while at the same time preserving a sense of nature or environment as the already-given *container* of such cultural practice. This would continue to have the reductive effect of picturing nature or environment as merely a mute and passive backdrop. The dwelling perspective instead involves a vision of nature and environment as active forces and participants in the unfolding of life, as both agents of change and that which is changed – as simultaneously both the object and subject of dwelling. In this way following Heidegger's injunction not to separate out being-in-the-world into discrete 'beings' and 'worlds', Ingold writes that

> from a phenomenological standpoint . . . the world emerges with its properties alongside the emergence of the perceiver as person,

against the background of involved activity. Since the person is a being-in-the-world, the coming-into-being of the person is part and parcel of the process of coming-into-being of the world as a whole.

(Ingold, 2000, p.168)

Here, what is especially pertinent is the manner in which Ingold closely aligns the notion of dwelling with *landscape* in particular. Indeed, in some ways he presents the two as indissolubly intertwined: for Ingold, landscape is dwelling. 'The temporality of the landscape' moves ineluctably towards this conclusion, initially by defining what landscape is *not*. Landscape is not, first of all, 'land', in the specific sense of an area or volume that can be mathematically described, quantified and measured. In contrast to mere land or acreage, landscape is here understood to be 'qualitative and heterogeneous' (ibid., p.190): it is not an amount of something, but a quality of feeling, in the end an emotional investment. Second, Ingold asserts, landscape is not 'nature', if by nature we specifically mean the Western notion of a realm external to human life and thought. If nature suggests externality and distance, landscape, by contrast is 'implicate' (ibid., p.191), both human and non-human, something with us, not against us. Lastly, Ingold draws a contrast between landscape and 'space', equating the latter with a cartographic representation of the earth's surface, one based upon a 'view from nowhere'. Landscape, again by contrast, is anchored in human, embodied perception.

In establishing these successive distinctions, Ingold steadily unfurls a phenomenological credo regarding landscape, one rooted in everyday, embodied dwelling. Landscape is first consanguineous with embodiment – an embodiment which, for Ingold as for Merleau-Ponty, is viewed as an ongoing generative process of acting in and with the world. In terms of dwelling, however, landscape is yet more closely entwined with *temporality*. Ingold (ibid., p.190), explicitly states that his consideration of landscape in tandem with temporality has the aim of 'dissolving the distinction between them'. To do so, he contrasts the abstract clock-time of chronology and history (as constructs of Western thought) with lived, social time; a temporality grounded in engaged, bodily activity, one in which we perceive time 'not as spectators, but as participants, in the very performance of our tasks' (ibid., p.196). Temporality, for Ingold, is something quintessentially performed, enacted. And so, in much the same way as an ocular, disengaged vision of landscape is replaced by the kinesthetic involvement of dwelling, the linear, universal time of history is replaced by a temporality phenomenologically grounded in lived, corporeal experiences. The landscape, both the milieu and the activity of dwelling, thus becomes ontologically saturated with temporality; the two are fused and indissoluble as a phenomenological 'whole' – 'the process of becoming of the world as a whole' (ibid., p.201).

At the heart of the landscape-as-dwelling thesis is an assertion of 'the fundamental indissolubility of the connections between persons and landscape' (ibid., p.55, original emphasis). Clearly, Ingold moves away from a mind/matter duality, such that landscape can nowise be divided up into first a set of immaterial symbolic meanings and second a 'physical landscape'. Instead, 'through living in it, the landscape becomes a part of us, just as we are part of it' (ibid., p.191). The dwelling perspective goes further than this, however, in dissolving distinctions between reflective thought and practical action, signification and performance. Ingold uses the term 'taskscape' to refer to the ongoing rhythms of everyday practice and performance, and his argument is that 'by re-placing the tasks of human dwelling in their proper context with the process of becoming of the world as a whole, we can do away with the dichotomy between taskscape and landscape' (ibid., p.201).

Thus, for Ingold (ibid., p.193), 'landscape is the world as it is known to those who dwell therein'. However, and crucially, it may also be defined as 'the everyday project of dwelling in the world' (ibid., p.191). Landscape is presented by Ingold as a *milieu of involvement*: it is neither a known and represented environment in or upon which meaningful human practice takes place, nor simply that practice itself. Landscape is *both* – both performative sensorium *and* site and source of cultural meaning and symbolism:

> telling a story is not like weaving a tapestry to *cover up* the world . . . [landscape] has both transparency and depth: transparency because one can see into it; depth, because the more one looks the further one sees. Far from dressing up a plain reality with layers of metaphor, or representing it, map-like, in the imagination, songs, stories and designs serve to conduct the attention of performers *into* the world. . . . At its most intense, the boundaries between person and place, or between the self and the landscape, dissolve altogether.
>
> (Ingold, 2000, p.56, original emphasis)

In conclusion, this phenomenology works to describe a landscape that is no longer a cultural frame, a 'way of seeing'. Nor is it a physical surface, an inert terrain. Landscape instead becomes the ongoing practice and process of dwelling. As Ingold summarises,

> The landscape, in short, is not a totality that you or anyone else can look *at*, it is rather the world *in* which we stand in taking up a point of view on our surroundings. And it is within the context of this attentive involvement in the landscape that the human imagination gets to work in fashioning ideas about it. For the landscape, to borrow a phrase from Merleau-Ponty, is not so much the object as 'the homeland of our thoughts'.
>
> (Ingold, 2000, emphasis in original)

## 5.4 'LANDSCAPING': PHENOMENOLOGY, NON-REPRESENTATIONAL THEORY AND PERFORMANCE

### 5.4.1 Non-representational theory

The phenomenological understanding of landscape advanced by Ingold raises a number of questions and can, of course, itself be subjected to a series of criticisms. However, I want to postpone that task until the next section. This is because I want to devote this section to discussing recent substantive work within cultural geography, material cultural studies, archaeology and anthropology that follows a broadly phenomenological approach. In addition, this discussion aims to provide something of a bridge between this chapter and Chapter 6 which follows. As noted above, amongst other things, Chapter 6 shall take a prospective look at the place of landscape within current and emerging agendas in cultural geography in particular. Here, in some ways laying the ground for those discussions, having already outlined in some detail Ingold's arguments, I will discuss more broadly and generally recent work, from cultural geography and beyond, that is attending to the embodied practice and performance of landscape.

The first and most crucial point to note here is that the take-up of Ingold's work within cultural geographies of landscape in particular has occurred primarily through the advent of 'non-representational theory', or what is sometimes called 'the performative turn'. First of all then I want to pay some attention to this movement, its inspirations and agendas. In general, what I am referring to is the fact that over the past ten years or so, and especially since about the year 2000, varied notions of performance, practice, materiality and embodied agency have increasingly come to the fore in human geographical research. Another way of putting this would be to say that there has been both a rhetorical and substantive shift, from studies of *representations* of landscape, nature, identity, space, place, the body and so on, to studies instead investigating various *performances* and performativities of these tropes.

The initial agendas of non-representational theory – and indeed the term itself – were advanced in a sequence of papers by its leading advocate, Nigel Thrift (1996, 1997, 1999), and Ingold's account of the dwelling perspective was a key citation for these papers. Echoing Ingold's arguments, Thrift sought first to criticise the constructivist epistemology of 'new' cultural geographies, with its focus upon the composition and communication of cultural meaning via text, sign, image and symbol. Thus, he wrote that the difficulty with metaphors such as 'landscape as text' (e.g. Barnes and Duncan, 1992), is that 'a hardly problematised sphere of representation is allowed to take precedence over lived experience and materiality' (Thrift, 1996, p.4). The argument is that, in the broadest sense, the art historical, literary and discursive sensibilities propagating across cultural geography in the early 1990s had, in tandem with a widely deployed cultural politics of identity, led to a situation in which everyday life, embodied experience and practice in general were considered as the secondary effects or outworkings of a more primary realm of cultural discourse and already structured social meaning.

For Thrift (2001; Thrift and Dewsbury, 2000), therefore, new cultural geographies had become '*dead geographies*'. Constructivist approaches and understandings drained the life out of things they studied. Human bodies, for instance, became blank slates awaiting the inscription of cultural meanings constructed at a putatively 'higher' discursive level (norms, values and beliefs regarding appropriate body shape, dress, behaviour, etc.). Nature became a social construct, a set of ideas about nature (for example the landscape way of seeing as a specifically Western means of structuring culture–nature relations). Human agency and identity became a cipher for the operation of a relatively fixed set of power relations operating via preformed social codes and categories: race, gender, class and sexuality conceived as sets of consistent inscriptive mechanisms.

In identifying these issues, and in formulating alternatives, geographers allying with and fashioning non-representational agendas have made use of a wide range of conceptual and philosophical resources, and are in turn, as will be discussed below and in the next chapter, developing an equally wide range of new topics, vocabularies, arguments and techniques. Non-representational theory, then, is not a singular thing. Rather, as Lorimer (2005, p.83) notes, this is 'an umbrella term for diverse work that seeks better to cope with our self-evidently more-than-human, more-than-textual, multisensual worlds'.

A first point to make in this regard is that non-representational theory does not mean *anti*-representational theory. In making this assumption, some commentators (e.g. Nash, 2000; Cresswell, 2002) have concluded that non-representational theory is concerned only with a realm of bodily habits, tics, routines and reflexes lying outside of both conscious thought and the shared social world of codes, norms and conventions. Lorimer's (2005) alternative phrasing, 'more-than-representational', perhaps better captures the agenda here. Non-representational theory, it has been argued, is concerned to develop new approaches to body *and* society, culture *and* nature, thought *and* action, representation *and* practice. Thus, as Dewsbury *et al.* (2002, p.438) write,

> non-representational theory is . . . characterised by a firm belief in the actuality of representation. It does not approach representations as masks, gazes, reflections, veils, dreams, ideologies, as anything, in short, that is a covering which is laid over the ontic. Non-representational theory takes representation seriously; representation not as a code to be broken or as a illusion to be dispelled rather representations are apprehended as performative in themselves; as doings. The point here is to redirect attention from the posited meaning towards the material compositions and conduct of representations.

In other words, the act of representing (speaking, painting, writing) is understood by non-representational theory to be *in* and *of* the world of embodied practice and performance, rather than taking place outside of that world, or being anterior to, and determinative of, that world. Or, to put this another way, the world is understood to be continually in the making – processual and performative – rather than stabilised or structured via messages in texts and images. As a consequence, therefore,

The focus falls on how life takes shape and gains expression in shared experiences, everyday routines, fleeting encounters, embodied move- ments, precognitive triggers, practical skills, affective intensities, enduring urges, unexceptional interactions and sensuous dispo- sitions. Attention to these kinds of expression, it is contended, offers an escape from the established academic habit of striving to uncover meanings and values that apparently await our discovery, inter- pretation, judgement and ultimate representation.

(Lorimer, 2005, p.84)

As should be clear from these statements, non-representational theory very much foregrounds the lived body and bodily practice and per- formance. Embodiment is here understood as both a substantive topic of study and as the inescapable medium in which sense is made and subjectivity is performed (see Harrison, 2000). In this way, therefore, phenomeno- logical conceptions of embodiment, particularly those associated with Merleau-Ponty, have been a significant source of inspiration for writing in this area (e.g. Wylie, 2002a, 2005; Cloke and Jones, 2001; Cresswell, 2003). But equally, a focus on bodily techniques and display in everyday life has drawn upon Judith Butler's (1993) well-known work on the performativity of identity, on ethnomethodology (e.g. Laurier and Philo, 2004) and on the rich and varied traditions of performance studies in theatre, music, drama and therapy (e.g. see Dewsbury, 2000; McCormack, 2002, 2003).

Thus, while non-representational theory recuperates and reinvigorates phenomenological accounts of embodiment, perception and human being- in-the-world, it is important to note that it is also pursuing intellectual trajectories that are in many ways quite opposed to phenomenological forms of understanding. One especially influential arc is a form of con- temporary vitalism drawing upon the work of Gilles Deleuze and Felix Guattari (1988, 1994; also Deleuze, 1992, 1994a; and Massumi, 2002). This is a vision of how the world works that has come strongly to the fore in current cultural geography, in so far as it affirmatively envisions a world continually in a process of relating, becoming, proliferating and differentiating. A vitalist ontology such as this in many ways countermands phenomenological perspectives, by advocating a radically non-subjective account of life, one in which the circulation of non-personal and non- human affects, forces and singularities is understood to supersede, rather than supplement, notions of being-in-the-world. Equally, current forms of relational materialism, devolving from actor-network theory (Latour, 1993; Law and Hassard, 1999; Pels *et al.*, 2002), and crucial to contemporary geographies of culture–nature in particular (e.g. Whatmore, 2002), are strongly anti-phenomenological and anti-humanist, picturing, as they do, a 'topological' vision of culture–nature relations from which terms such as landscape and subjectivity are often strikingly absent. Again, the growing influence of these intellectual currents, and their potential impact on the theorisation and study of landscape, will be further explored in the next chapter.

### 5.4.2 The body, practice and touch

In terms of landscape specifically, however, the advent of phenomenological and non-representational theories of embodied practice and performance *has* occasioned a significant shift of emphasis. One quick way of capturing the essence of this shift would be to speak about a move from 'images of landscape' to 'landscaping'. In other words, whereas a previous generation had focused upon already-made *representations* of landscape (texts, images) and their varied negotiation of cultural discourse and regimes of power, current work more commonly turns towards *practices* of landscape and, especially, towards the simultaneous and ongoing shaping of self, body and landscape via practice and performance.

When I speak of 'practices' here I am referring above all to common, embodied cultural practices – not to esoteric or mysterious activities, but rather to familiar and recognisable things such as walking, looking, driving, cycling, climbing and gardening. The implicit argument behind many recent cultural geographies of landscape is that such practices may be understood as 'embodied acts of landscaping' (Lorimer, 2005, p.85), in which self, landscape and indeed culture itself inhere, circulate and emerge. This further involves a broad methodological shift, moving from a largely interpretative and discursive standpoint towards a more ethnographic and performative ethos. In this way, attentive analysis of, and, quite often, direct personal *participation* in, embodied acts of landscaping becomes the substantive task for contemporary landscape studies.

An especially notable feature of recent landscape work has been the increased attention paid to *tactile*, as opposed to *visual*, landscape experiences. The conceptual shift from landscape-as-image to landscape-as-dwelling correlates with a substantive shift from *horizon* to *earth*. In general, the proliferation of research on the body and embodied experience turns landscape from a distant object or spectacle to be visually surveyed to an up-close, intimate and proximate material milieu of engagement and practice. Landscape becomes the close-at-hand, that which is both touching and touched, an affective handling through which self and world emerge and entwine. In this way also notions of *landscape* begin to merge with notions of *place*; landscape and place conjoin intimacy, locality and tactile inhabitation. Kevin Hetherington's (2003, pp.1,936–1,937) discussion of touch and place summarises this valuing of tactility well:

> We think of touch, at least initially, as up close, local, and specific (proximal) in its way of knowing. It is also inherently dialogical in character. We are often touched by what we touch: a lover's body, a child's hand ... or a sculpture, or something more mundane, a set of rosary beads in a pocket perhaps (but it could equally be the familiarity of our keys), or a vase on a shelf. ... Praesentia [encounter through touch] is a way of knowing the world that is both inside and outside knowledge as a set of representational practices. It is also performative and generative of knowledge communicated other than through representation. Both a form of the present and a form of presencing something absent, it can be found in tacitly skilled, haptic reaching out and does not presume in advance the necessity of an engagement in the act of visual representation, let alone its outcome as knowledge that can be communicated discursively to others. Rather, praesentia presumes only an involvement and a confirmation of subject formation in the materiality of the world.

'Subject formation in the materiality of the world' is a key theme for much recent landscape writing, developing the broadly phenomenological argument that 'humans do not act as subjects in an object world but are constituted as perceiving beings at the interface between subject and object' (ibid., p.1,938). In consequence there are now numerous studies of up-close engagement, for example Cloke and Jones's (2001, p.652) study of orchard-growing in Somerset, which explicitly adopts Ingold's dwelling perspective to 'account for the intimate, rich, intense, making of the world, where networks fold and form and interact in particular formations which include what we know as "places"'. Here, the sensuous and tactile experience of generation and seasonality testifies to 'how human actants are embedded in landscapes, how nature and culture are bound together, and how land-scape invariably has time-depth which relates the present to past futures and future pasts' (ibid., p.664). These sorts of arguments are echoed in Crouch's (2003) work on allotment gardening as 'grounded performance', or a 'tangle in the mundane', and, with a different emphasis upon non-human agency, in Degen et al.'s (2006, forthcoming) study of 'passionate involvements' with vegetable growing and knotweed clearance schemes. Crouch relates the seemingly humdrum activity of gardening to deeply felt senses of self, landscape and nature, such that for one of his respondents gardening was:

> how she spaces herself; her body engages in intimacy of space, sig-nificantly on her own terms, with movements amongst multisensual encounters. Carol touches, bends, and kneels; she moves her body and spaces the gaps, and their objects, between vegetation, earth, insects, the air, and herself. She finds her feeling of life through what her body does there.
>
> (Crouch, 2003, p.1,953)

Moving to a maybe more extreme tactile realm, my own work (Wylie, 2002b, c), on Scott's and Amundsen's journeys to the South Pole between 1910 and 1912, also aims to foreground the tactile, or more broadly 'haptic' (a term which conveys the sense that vision may also be a form of touch), as a key register through which self and landscape are produced. Defining landscape as 'a concrete and sensuous concatenation of material forces' (2002b, p.251) through and against which the British and Norwegian polar parties defined themselves, this work sought to examine their '[bodily] comportments and competencies, not as reliefs upon a polar environment viewed as a sublime background, but as ways of becoming polar forged through intimacies with that environment' (ibid.). It thus tried to describe in particular,

> the caring for the self which the materiality of the Antarctic landscape demands. Each night in the tents, eyes blinded by the luminous intimacy of the landscape are treated with zinc sulphate and cocaine, then swaddled in rags and tea leaves. Frozen feet and hands are placed upon the warm chests and stomachs of consenting com-panions in a series of awkward embraces, unlikely arrangements of

bodily parts. Antarctica demands, above all, that the frontiers of one's body be rigorously established and maintained.

(Wylie, 2002b, p.259)

In a slightly different sense, Lewis's (2000) account of adventure climbing wants to proclaim tactility, immersion and engagement as emancipatory experiences. For the climber, 'freedom becomes a form of embodied awareness: a choosing to sense and, more specifically, a choosing to *feel* and *touch* an environment' (2000, p.58, original emphasis), and so, 'as an extreme experience, climbing becomes a kind of corporeal subversive politics ripe with possibility for renewal that feeds back into private and social life, inflecting it with new horizons for human embodied agency' (ibid., p.63). This type of explicit politicising of bodily practice has become quite rare, however, and for the most part recent cultural geographical work on embodied acts of landscaping segues less with critical sociologies and more with new forms of material culture studies regarding relations with the object-world (e.g. see DeSilvey (2003) and Hitchings's (2003) work on gardening).

### 5.4.3 Landscape, material culture and archaeology

Emerging as a more distinctively defined and yet diversifying field over the past twenty years (e.g. compare Miller (1997) and Miller (2005)), material culture studies examines the relations between people and things, and thus tends to position practical engagements with crafted and significant objects as a primary activity through which sense and value are made and communicated. Anthropology and archaeology have been the two disciplines most clearly associated with a material cultures perspective, perhaps obviously in the case of the latter, which takes a realm of recovered objects and artefacts as its primary domain, and, classically, aims to 'reconstruct' past cultures from their material remnants. And one strand of material culture studies in which issues of *landscape* are particularly on the agenda is an emerging area of overlap between archaeology, phenomenology and performance studies (Shanks and Pearson, 2001). Like human geography, via an encounter with elements of cultural and critical theory, and more recently phenomenology (see Shanks and Tilley, 1992; Thomas, 2004; Tilley, 1994, 2004), some elements of archaeological practice have gradually disentangled themselves from their empirical, descriptive and classificatory roots. In this way, the body and the senses have been fore-grounded in archaeological work focusing upon direct encounters with, and performative engagements of, material landscapes and artefacts (sites, buildings, standing stones) (see Bruck, 2005; Hamilakis *et al.*, 2002). Here, corporeal phenomenology emerges strongly as a particular form of research practice. Michael Shanks and Mike Pearson's (2001) co-authored text *Theatre/Archaeology* is a striking example of this work. 'We look, we listen, we touch,' Shanks and Pearson (2001, p.134) argue, 'we begin to inhabit and measure this world through our sensory experience of it.' Pursuing this principle, *Theatre/Archaeology* stages (and the metaphor is apt) a series of experimental encounters with sites, narratives and theoretical legacies – encounters *through which* senses of self and landscape are conducted and elaborated. For the authors, this is

> a science/fiction, a mixture of narration and scientific practices, an integrated approach to recording, writing and illustrating the material past. Here archaeology and performance are jointly active in mobilising the past, in making creative use of its various fragments in forging cultural memory out of varied interests and remains, in developing cultural ecologies (relating different fields of social and personal experience in the context of varied and contradictory interests).
>
> (Shanks and Pearson, 2001, p.131)

Here, landscape is activated and mobilised within a complex *mélange* of elements, purposefully intended to scramble conventional orders and disciplinary codes. Thus,

> In theatre/archaeology documents, ruins and traces are reconstituted as real-time events. But we do not present theory, method and case study separately. . . . This is a combination of performed material, narratives and ruminations on the theory and practice of theatre/archaeology. At pivotal points we consider concepts of landscape, temporality, interpenetration, evocation as an oblique strategy of representation, site-specific theatre, story-telling, the guided visit, deep mapping, memory and identity – those practices which help constitute both city and country.
>
> (Shanks and Pearson, 2001, p.131)

The work of the archaeologist Christopher Tilley (1994, 2004), while aligned with these sorts of experimental engagements, elaborates phenomenological practice in what could be seen as a more purist fashion. Tilley's approach to landscape is strongly tactile, field-oriented and site-specific. In particular, in *The Materiality of Stone* (2004), he presents the results of a series of direct, embodied encounters – as he puts it, 'empirical without being empiricist' (p.219) – with prehistoric stone artefacts: Breton standing stones, Maltese temples and Swedish rock carvings and cairns. The argument is that such encounters not only open up a realm of practical, sensuous knowing that standard archaeological representation of past landscapes (in the form of photos, models, diagrams, maps) forecloses, they also open the past itself:

> archaeological excavation results in an absolute closure of the past and the reports only provide limited possibilities for further understandings and reinterpretations, because all that is left is a representation of that which has been disturbed, at best, or destroyed.
>
> (Tilley, 2004, p.219)

The ethos of performative landscape archaeology thus echoes non-representational theory's concern that the conversion of life, matter and practice into text, sign and image produces only 'dead geographies' (Thrift and Dewsbury, 2000). An especially notable feature of Tilley's argument is that he includes Ingold's (2000) work amongst a long list of writers castigated for reducing the bodily experience of landscape to a matter of cultural and most commonly visual representation. This is because Ingold's account of landscape and temporality uses a *painting* – Brueghel's *The Harvesters* – as a vehicle for elaborating notions of dwelling. This does have the odd effect of inferring dwelling practices from a representation of dwelling, but the example must be set in the context of Ingold's wider point that landscape images and narratives conduct performers *into* the world – that representation is a form of practice, in effect. For Tilley, however, a strict adherence to the precepts of Merleau-Ponty's carnal philosophy is a prerequisite for a properly phenomenological approach to landscape and place. Thus, he argues, 'any study begins with lived experience, being there, in the world. It must necessarily be embodied, centred in a body opening itself out to the world, a carnal relationship' (2004, p.29). And, elaborated out from primary bodily movement and perception, Tilley seeks to reclaim landscape as a holistic term, larger than place – a term that gathers together body, place, perception and relationships between people and between people and things:

> Experience of the world always extends from the body and expands beyond the particularities of place. A more holistic perspective is required, one that links bodies, movements and places into a whole, and this is why the term 'landscape' has utility. . . . Landscapes can be most parsimoniously defined as perceived and embodied sets of relationships between places, a structure of human feeling, emotion, dwelling, movement and practical activity.
>
> (Tilley, 2004, p.25)

---

**Box 5.2 DEVON MEMORIES**

In the summer of the year 2000 I spent three weeks on holiday in Devon, in south-west England. We stayed in a house in a very rural, even isolated location; in east Devon, not far from the county border with Somerset, deep in the folds of the broad strip of rolling countryside which divides the uplands of Exmoor, to the north, and Dartmoor, to the south. This house stood high upon one side of the valley of the river Lowman, a few miles to the north-east of Tiverton, the nearest town. During those three weeks, amongst other things, I read a book called *The Imperative* (1998) by the American phenomenologist Alphonso Lingis, and so gradually came to a different understanding of landscape.

I knew nothing at all about the history of the locality we found ourselves in, the curving valley which the house overlooked. One notable fact, however, was that, along with large tracts of the surrounding countryside, the valley was owned and administered by a distant trust in a quasi-feudal manner. This accounted for the almost total lack of visible development in the area. Apart from the house in which we were staying, there were only three tenanted farmhouses within a radius of about three miles; the rolling land was mostly pasture with occasional mature copses, and, down by

---

the river, wild flower meadows. It was absurdly picturesque, it demanded to be seen as naturally beautiful and aesthetically satisfying in a way that years of lectures and reading had told me was an ideological façade.

Although several main roads were only a few miles away, there was hardly any traffic. The valley was unmarked on tourist maps and the few twisting, hedged lanes which ran through it led to nowhere in particular, frequently turning back on themselves. Electricity entered the valley along a string of skewed wooden pylons, often lost amidst the greenery, and the supply was a little unreliable. Isolation and insularity was married with extreme tranquillity: no people, few sounds.

In this idyll, high summer, cloudless skies, the house on its eminence opened onto a broad and softly falling scene. The view was serene: shades of green and yellow in pastel-patterns dappling into the distance. I saw enclosed fields of crop and pasture, small herds of grazing animals, sinuous hedge-thickened lanes. But not a single building. The lone line of wooden pylons stalked away down the valley from the house, into a distant, dozy pastoral haze. The valley was that of a youthful river. On each side of the house save in front – on the edges and frames of vision – slopes curved up and around, their angle and gather shaping a rounded fullness that did not terminate sharply in any summit or ridge, but rather passed into depth, into further, indefinite folds and rises invisible from the house, yet present as a sort of character, an undulating topography which girdled the particular point of view, and lent it assurance. From the house, looking forward as it did, the ground fell and rolled gently away toward the valley floor. This view falling away and this encircling topography together configured a rich depth, a satisfying set of distances and contours. Everything seemed settled and calibrated, cupped in a steady hand. In the centre of the view from the house a large, old oak partly obscured the point of the middle distance, the place where the valley printed its terminating 'v'. Beyond, a distant vista of recumbent fields extended indefinitely away.

continued

This view sustained me as I sat reading in the garden, and I started to feel that the house's situation could not have been accidental, its outlook once upon a time surprising those who looked up from laying foundations, and now forming just a pleasant addition. But it was difficult to account for its qualities.

Thinking of landscape as a way of seeing – as a culturally specific way of looking at and picturing the world – would provide one means of understanding and writing about this Devonian pastoral. Beyond all else, the art evolving from this way of seeing has had a very important role in forming European tastes for certain types of visual landscape. The graduated, bucolic perspective before me, for instance, which seemed to have already organised itself into a fore, middle and background, confirmed the tenets of classical landscape art: order, balance, harmony. And then there were several centuries' worth of rural imagery and myth surreptitiously at work as I was watching, so to speak – notions of the countryside and of nature as wholesome, harmonious, authentic, enriching, restorative and so on.

The landscape way of seeing is also as much about *viewer* as *view*, of course. More precisely it configures a viewer, a view, and a set of relationships between them. I was being cast in a starring role: the visual observer of the landscape, rapt but distanced, a privileged witness in various ways. The gaze locates perception, knowledge and imagination primarily within the observing subject. This way of seeing establishes certain versions of the inside and the outside, observer and observed, as at once aesthetically compelling *and* intellectually authoritative.

But reading *The Imperative* (Lingis, 1998), there and then, made me question key elements of this account – the account of landscape as a visual ideology. Its author, Alphonso Lingis, is a phenomenologist inspired by Merleau-Ponty and by Emmanuel Levinas, but also a distinctive writer: his books are an odd blend of phenomenological discussion, ethics and forms of travel writing. Within a suite of books also including *Foreign Bodies* (1994) and *Dangerous Emotions* (1999), *The Imperative* is the closest Lingis has come to a programmatic account of perception, materiality and subjectivity.

What I took from *The Imperative*, on first reading, was a sense that the *materiality* of landscape was highly significant. Here, materiality refers to something more than the landscape's bare physical presence and solidity, its there-ness and matter-of-factness. Nor, for Lingis, does speaking of landscape's materiality involve a critical or Marxian narrative centred on the material production of landscape through industrial and agricultural processes, and the concomitant material transformation of human lives (Chapter 4, Section 4.3). Instead, working out from a corporeal, phenomenological perspective, *The Imperative* paints a picture of landscape's materiality in terms of force, animation and perception. The external, material landscape we encounter is not mere lumpen or dead matter, a lifeless mass onto which we project meaning and value. Instead,

> as soon as we open our eyes, the light, the depth of the tangible, the hum of the environment besets us, soliciting, enticing, badgering. The exterior is not an empty and neutral but a resplendent, beguiling, bleak or stifling expanse. Reality weighs on us; we cannot be indifferent to it.
>
> (Lingis, 1998, p.119)

Another way of putting this is to say that:

> it is not that things barely show themselves, behind illusory appearances fabricated by our subjectivity; it is that things are exorbitantly exhibitionist. *The landscape resounds*; facades, caricatures, halos, shadows dance across it. Under the sunlight extends the pageantry of things.
>
> (*ibid.*, p.100, my emphasis)

In other words, *The Imperative* presents the matter of landscape as something already alive, active, animate. Lingis's purpose in *The*

continued

*Imperative* is to supply 'a phenomenology of perception which brings out the order and ordinance inherent in the perceived field' (*ibid.*, p.5). The focus is thus not upon the perception of matter by an already-given, perceiving subject; instead from the first we detect, in materialities and liquidities, in telluric, celestial and organic processes, *imperatives to action* which guide, imply and ordain corporeal sensibilities.

For Lingis, therefore, the visible world not only transcends the subject, in the sense of exceeding our perception of it, it also summons and directs it in certain ways. A sense of self arises in the course of gazing as a vector of response to exteriorities – to encountered others, to sights and sounds, to both textures and intangibilities. This is because 'the layout of the field of perception functions as an ordinance imperative for perceptual exploration' (*ibid.*, p.215). In this way, as an animate matter, intertwined with our gaze upon it, landscape is more than a way of seeing. Landscape is that with which we see, a perception-with-the-world.

It was this sense of landscape that I first took from *The Imperative*, that summer in Devon. I could see, watching the valley in sunlight from the garden, that there was an agency and an affordance in the landscape itself, a certain animation that exceeded everything that could be written about it. Discursive legacies such as the picturesque and the pastoral of course interwove with this animacy, but they did not wholly exhaust it. The relation between self and landscape is not always or strictly that of observer and observed. The eyes of the gazing subject are not an exercise of judgement or a bestowal of meaning upon a passive and neutral scene. Instead these eyes arise and look through a tension with the world, with visibilities, sonorities and tangibilities, a unfolding tension that, as Lingis says, 'organises as it proceeds' (*ibid.*, p.31).

## 5.4.4 Landscape, mobility and embodied vision

As this important quote suggests, a focus upon close, tactile engagement in recent landscape phenomenologies is accompanied by a renewed attentiveness to issues of *movement* and *mobility* in landscape. This 'mobility', it should be noted, is not confined to physical travel. Instead, the adoption of the lived body as both domain of enquiry and research conduit has focused attention, in various ways, upon *perception-in-motion*. In this way landscape ceases to be understood as a static, framed gaze, and becomes instead the very interconnectivity of eye, body and land, a constantly emergent perceptual and material milieu. Thus, much writing in this area concerns itself with the experience of mobility, with being and becoming mobile as a particular trajectory of subjective cohesion *and* dissolution. For example, Spinney's (2006) work on cycling attends to the kinetic assemblage of self via bodily performance, technology and landscape. Here, as in other work in this area, there is a attempt to segue between phenomenological and discursive approaches to landscape and subjectivity, so as to enable a dual focus upon both bodily experience *and* the material, governmental frameworks that produce and regulate such experience (e.g. in Merriman's (2005a) work on driving practices). In a slightly different sense, mobility might be appropriated as a deliberate vehicle for opposing and transcending notions of landscape as static and sedentary. This is Tim Cresswell's (2003) argument in a paper celebrating J.B. Jackson's 'mobile view of landscape' (p.275) as a means of escaping and subverting official norms, and accessing an at once richer and more quotidian understanding of landscape as a mobile form of everyday lived practice.

With Jackson's mobile observer in mind, it is important to note that in a turn towards corporeality and performance there is a risk of unwittingly thinking that vision somehow 'belongs' to the mind, consciousness, discourse and so on. The outcome would be accounts assuming embodied experience meant non-visual experience. However, as Merleau-Ponty's writing demonstrated in depth (see Section 5.2), vision is of course as corporeal as the other senses, even if it has culturally and historically been associated with detached thought and 'mindful' knowledge. Taking Merleau-Ponty's work forward, a number of articles have explored vision-in-motion in particular. Again, my own piece on ascending Glastonbury Tor, a prominent hill in Somerset (Wylie, 2002a), is an exploration which explicitly seeks to describe vision in terms of intertwinings of self and world. Walking to the top of the Tor, the paper argues, is less about attaining a masterful perspective than about experiencing 'a certain involution or folding of the landscape as a whole. . . . The entire voyage to the summit is conducted through passages of visibility from which emerge points of view and fields of vision' (2002a, p.450). To speak of vision, of gazing upon landscape, is, following Merleau-Ponty, to speak about an intertwining through which observer and observed are assembled as such. Thus, the conclusion of this paper argued that:

> the voyage to the summit of the Tor involves interrogating and sometimes abandoning any distinction between vision and visible, seeing and thing seen. The Tor is never an object for a gaze, nor a viewpoint on its own. It is a modulation of the visible world which lets there be a gaze to behold things open and hidden. . . . As bodies we are caught in this world, this ontology of visibility, and like the Tor we both see and are seen. So it is through being embodied that we can see the Tor, climb it, see from it. The trip to the top is a voyage in which we are shaped by visible depth to be its witnesses. And from the summit of the Tor I can see that, as Merleau-Ponty (1962, p.181) says, 'my power of imagining is nothing but the persistence of the world around me'.
>
> (Wylie, 2002a, p.454)

Several other articles have begun to explore the nexus of materiality, corporeality and perception. For example, Dubow's (2001) historical analysis of travel, first by sea to the South African Cape region and thence by cart through the interior, moves away from the discursive and structural readings of colonial travel (e.g. Blunt, 1994; Phillips, 1997) discussed in Chapter 4, Section 4.5, to engage instead with the materialities of vision – materialities through which South Africa takes shape as territory. This emphasis upon the direct and uneliminable role of corporeal performance in the production of knowledge also surfaces in Foster's (1998) account of the writer John Buchan's travels in southern Africa, and in Lorimer and Lund's (2003) account of how practices of walking, hiking and visualising – 'pedestrian geographies' – are assembled into expert knowings of Scotland's mountains. And finally in this regard, Ingold (2005) himself has commented upon the mobility of the sky itself, and its role in perception, making the strikingly obvious but largely overlooked point (but see Brassley,

1998), that 'in the scholarly literature on visual perception, scarcely a word is to be found on the question of how the weather impacts upon practices of vision' (2005, p.97, original emphasis).

With this invocation of the elemental, we come to a final, as-yet-nascent theme in the ongoing encounter between landscape and phenomenology. Dwelling, as described by Heidegger (1996), has a mythopoetic or cosmological dimension. It goes beyond a description of lived experience, and beyond even 'ontology' – the study of human being and existence – to invoke a more primordial, flickering and obscure alliance of earth, sky, divinity and mortality. Dwelling is, finally, a poetic vision of the gathering together of earth and humanity as landscape. This, of course, is problematic in many respects, and the next section will examine in more detail the potential critical errors into which unwary users of the dwelling perspective can fall. And it is also easy, even sometimes satisfying, to scoff at Heidegger's solemnity and portentousness. Nevertheless it is the case that because of this poetic, metaphysical aspect of dwelling, and because phenomenology in general is opposed to the standard objective, realist and empiricist tenor of much Western thought, culture and science, some writers have found in this tradition a decidedly spiritual tone. For Abrams (1996), for example, phenomenology holds the key to recovering a more authentic, holistic and ecologically balanced relationship with the earth. More prosaically, perhaps, forms of landscape phenomenology may be deployed in the analysis of alternative, 'new age', spiritual and therapeutic relationships between self, body and nature. Holloway's (2003) use of Merleau-Ponty's work to approach questions of spiritual practice in and around the Glastonbury landscape is a good example of this type of work and, more generally perhaps, work on 'therapeutic landscapes' (Gesler and Kearns, 1998) could fall into this category, emphasising as it does holistic connections between landscape qualities, well-being and health. While issues of embodiment and performance have been somewhat neglected in this sub-field, David Conradson (2005, p.338) has opened up the argument that 'the therapeutic landscape experience is best approached as a relational outcome, as something that emerges through a complex set of transactions between a person and their broader socio-environmental setting'. The use of various therapeutic practices and techniques designed to provoke alternative experiences of self, time and nature (see Thrift, 2000b) is perhaps an area into which the insights and procedures of landscape phenomenologies may be usefully extended.

## 5.5 CRITIQUES OF LANDSCAPE PHENOMENOLOGY

As a mode of enquiry, phenomenology has always been subject to criticism, and has even, it can be suggested, been viewed with suspicion by some geographers. Therefore, despite the range of substantive recent research discussed in the previous section, phenomenology's position as a legitimate approach to landscape remains, arguably, precarious. Particular antipathy to phenomenological approaches is often evident in the case of writers whose work is located within critical, radical and Marxist traditions. For example, in his book-length account of the geographies of nature (and thereof, the nature of geography), Castree (2005) does not mention at any point the extensive corpus of phenomenological writing on culture–nature relations discussed above.

In this section I want to outline some of the reasons why phenomenology has been, and continues to be, viewed by many as a problematic endeavour. First of all, then, with reference to critical, radical and Marxist approaches, phenomenology is often negatively understood as an 'individualist' philosophy, one ineluctably emphasising the actions, values, emotions and perceptions of individuals. The problem, from a critical and/or Marxist perspective, is that this 'people-centred' approach (and phenomenological or humanistic geographies are often described in the approaches anthologies as 'people-centred', e.g. see Cloke *et al.*, 1991; Aitken and Valentine, 2006), insufficiently conceptualises and addresses the varied social, economic, historical and political contexts in which individuality is, the argument runs, primarily shaped and determined. Phenomenological approaches, and in particular the early forms of geographical phenomenology epitomised by writers such as Yi-Fu Tuan, Edward Relph and David Seamon, thus stand condemned as non-critical celebrations of human individuality. Despite the fact that varied phenomenological writers – and here I would include elements of the work of the very influential French sociologist Pierre Bourdieu (1977, 1984) – have emphasised the quintessentially social and intersubjective character of individual life, phenomenology is viewed as lacking the critical purchase provided by the argument that historical and material circumstances hold the key to understanding both individual and social worlds. To put this another way, it is argued that attending to individual human emotion and perceptions inevitably reifies such emotion and perception, treating it as a priori and given, and thus failing to recognise that the very notion of the free, autonomous individual is to some degree an ideological fabrication essential to the functioning of a capitalist socio-economic system. Thus, phenomenology does not supply the armature requisite to the project of a critical social science, one intent upon a critical assessment of, and reshaping of, current social, political and economic arrangements. In this way, not only radical and Marxist but more broadly critical, Foucauldian and feminist geographers have often fought shy of phenomenology. For example, Catherine Nash (2000, p.660) voices what is perhaps a widespread feeling in commenting that,

> My wariness about abstract accounts of body practices and the return to phenomenological notions of 'being-in-the-world' arises also from the danger . . . that they constitute a retreat from feminism and the politics of the body in favour of the individualistic and universalizing sovereign subject.

In the most general terms, then, various forms of critical theorisation emphasising the role of social and economic structures, and the broader politics of identity construction have had difficulties with a phenomenological approach, because such an approach appears to neglect the constraining and determining effects of forms of *power*. Much cultural geographical writing on landscape, of course, as was discussed through Chapters 3 and 4, explicitly aligns itself with such critical theorisation, and seeks to interrogate landscapes as expressions of cultural, economic and political power, and so, though it would be wrong to overemphasise this, there remains something of a tension between such work and that, like Ingold's (2000) or Tilley's (2004), which follows a perhaps more classically phenomenological line.

If focusing upon individual agency at the expense of theorising the role of historical and material contexts has been one major charge levelled against phenomenology, another is that the approach is tainted by a certain *romanticism*. Such romanticism is most commonly articulated through an implicit valuing of rural, pre-modern and non-Western ways of life. This criticism can, in fact, be directed against Ingold's (2000) work on phenomenology, landscape and dwelling. As Cloke and Jones (2001, p.661) note, Heidegger's conception of dwelling is at least partly rooted in a 'sinister . . . rustic romanticism'. Dwelling thus tends to be presented and understood in terms of a fictitious, pre-industrial, rural *gemeinschaft*, leading to a view of 'authentic landscapes, or communities, as consisting of diminishing pockets of harmonious, authentic dwelling in an ever-encroaching sea of alienation' (*ibid.*, p.657). In these terms, dwelling as a concept is at best hamstrung by nostalgia; at worst it installs and maintains a false boundary between the 'traditional' and the 'modern'.

Without explicitly confronting this difficulty, Ingold's 'The temporality of the landscape' attempts to sidestep it via reference to a cyclical, almost seasonal temporality, one which is 'essentially rhythmic' (Ingold, 2001, p.196). And these ideas of rhythm and seasonality are endorsed also by Cloke and Jones (2001, p.658), in their advocacy of dwelling 'in a dynamic, time-embedded sense, rather than in comparison with any fixed time-point referencing'. Yet it can be argued a 'rhythmic' temporality may also be quite problematic when presented as an essential element of dwelling. It too beckons us towards a certain vision of simple rural toil. This is especially evident when Cloke and Jones (*ibid.*, p.654, emphasis in original) speak of dwelling in terms of 'time-deepened experience . . . the experience of rootedness, the richness of things together *over time*'. The concept of dwelling is thus not only grounded within a rural, cyclical temporality, it would seem to be dependent upon such a temporality, and can only be achieved through it.

These arguments can be usefully scoped out further. As we have discussed in this chapter, phenomenology is an anti-Cartesian philosophy; it regards the divisions of mind and body, subject and object, observer and observed as secondary constructs, and seeks to re-establish a more primary, original engagement between self and world. Thus, at the start of his introduction to the *Phenomenology of Perception* (1962, p.xi) Maurice Merleau-Ponty argues that 'all of its efforts are directed towards re-establishing a direct and primitive contact with the world'. And the key word here is 'primitive'. It can be argued that phenomenology has, from its inception, been haunted by nostalgia – nostalgia for a supposedly more authentic, engaged and 'natural' perception of the world, one which we have, so the argument goes, lost as a result of the installation of an objective, modern, detached perspective. Phenomenology originally takes shape as a quest for lost essences and ultimate foundations. And so, in order to find itself, it has to look to the *long ago* and the *faraway* – to the pre-modern and the non-Western. In this sense it is probably no coincidence that two of the disciplines in which phenomenological approaches currently have most purchase are archaeology and anthropology. Archaeology, of course, is predominantly concerned with the understanding of past cultures and societies – the long ago – while anthropology has traditionally focused upon the study of non-Western cultures – the faraway.

The difficulty here is that phenomenological approaches run the risk of romanticising the pre-modern, and particularly the non-Western, in a manner that inadvertently perpetuates notions of the historical pre-eminence and priority of Western cultures. To argue that the beliefs and practices of non-Western cultures today can be equated with a pre-modern or pre-Cartesian perception of the world – to equate the faraway with the long ago – is to implicitly subscribe to a Eurocentric narrative in which human cultures are measured and known according to their place along the particular developmental path prescribed by the history of Europe. Similarly, to argue that non-Western peoples, in particular indigenous peoples, live their lives through a perception of the world which is more authentic and natural than that of the contemporary West is to project onto such peoples a latter-day version of the romantic fantasies of Arcadian innocence and oneness with nature which characterised many colonial and imperial representations of non-European others (as was discussed in Chapter 4, Section 4.5). In sum, some incarnations of phenomenology would seem to valorise and depend upon a certain vision of pre-modern, pre-technological conditions, and to romanticise relationships with land-scape that are non-objective and non-rational. A myth of natural, 'raw', or 'primitive' experience – one of the defining mythologies of Western self-consciousness – potentially finds an echo in phenomenological writing. Of course, writers such as Ingold (2000) and Tilley (2004) are well aware of this issue, and their work demonstrates that it can be successfully negotiated, but in less careful hands – for example Abram's (1996) attempt to blend phenomenology, ecology and new age spirituality – the problem is glaring. The philosopher Gilles Deleuze summarises the issue simply: 'what phenomenology sets up as a norm is "natural perception" and its conditions' (Deleuze, 1988a, p.57).

Mention of Deleuze here leads to a third and final set of criticisms of phenomenological approaches. Much of the early work of the leading lights of poststructural thought, for example Gilles Deleuze, Jacques Derrida and Michel Foucault, can be understood as an attempt to critique and sup-plant the prevailing post-war orthodoxy of existential phenomenology in Continental thought. In particular, through the 1950s and 1960s, in a conflict inaugurated by the philosopher Jean-Paul Sartre's famous state-ment that 'existentialism is a humanism', an opposition existed between a phenomenological approach emphasising agency, perception and subjectivity, and a structuralist approach alternatively stressing the central role of broader linguistic, symbolic and discursive structures in the fabrication of the modern notion of the speaking, perceiving subject. In a sense, much of what we know as 'poststructuralism' devolves to us from this opposition, and in turn presents a strong critique of the humanist and subjectivist tendencies of phenomenologies. Thus, texts such as Derrida's (1973) *Speech and Phenomena*, Deleuze's (1990) *The Logic of Sense* and Foucault's (1977) *The Order of Things*, all written the late 1960s, in many ways constitute an anti-phenomenology, attacking in turn its central notions of presence, perception and subjectivity.

For Foucault in particular, the problem of phenomenology is its insistence on taking the human subject as the measure of all things. Phenomenology continues to be the domain of those 'who refuse to think without immediately thinking that it is man who is thinking' (Foucault, 1977, pp.342–343). For example, in the case of Merleau-Ponty, even if the trajectory of his thinking carries him towards an exterior world of differences and depths which assembles the human subject, he still assumes that subjective experience of the world is what, ultimately, requires description. And in this way phenomenology continues to posit and depend upon a 'natural' perceiving subject, a pre-linguistic, pre-cultural lived body possessed of inherent perceptual arrangements and faculties: to repeat, 'what phenomenology sets up as a norm is "natural perception" and its conditions' (Deleuze, 1988a, p.57). Thus phenomenology involves a descent to a 'natural' strata of being or a search for 'an originary experience, a first complicity with the world which for us would form the basis of being able to speak about it' (*ibid.*, p.55). And thus also phenomenology imputes a pre-given perceiving subject *to whom* the world is given, a 'judging' subject who synthesises experience.

In this context, it is worth recalling Tim Ingold's summative definition of landscape:

> landscape, in short, is not a totality that you or anyone else can look *at*, it is rather the world *in* which we stand in taking up a point of view on our surroundings. And it is within the context of this attentive involvement in the landscape that the human imagination gets to work in fashioning ideas about it.
>
> (Ingold, 2000, p.207, emphasis in original)

The term 'attentive involvement' is telling here. In conceiving of the dwelling body-subject as an attuned accumulation of environmental aptitudes, Ingold's writing to an extent shares the problems of classical phenomenology, in particular the intentionalist sense that the lived body possesses 'natural' capacities to synthesise, polarise and organise the perceptual field. Thus there occurs even in Ingold's work a partial reintroduction of the intentional subject that so much poststructural theory – and so much work on landscape by cultural geographers – sought to disassemble. In other words, Ingold's work perhaps remains too subject-centered, too humanist even, in so far as it tends to *replace* a detached meaning-bestowing 'cultural' mind with an active, sturdy and involved dwelling body.

## 5.6 CONCLUSION

This chapter has discussed in depth the comparatively recent re-emergence of phenomenological approaches to landscape in cultural geography, anthropology and interpretative archaeology. In facing this task, given that this is a relatively new and in some ways complex area of work, the chapter has also sought to spell out some principles of a phenomenological approach. Thus a substantive section of the chapter (5.2) was devoted to a discussion, via the work of Descartes and Merleau-Ponty, of how a phenomenological approach to landscape involves moving conceptually from a disembodied and detached gaze to an engaged and lived body. Following this, Tim Ingold's writing on embodiment, perception and dwelling (5.3) merited close attention as an important and influential account of landscape post its definition as a 'way of seeing' or 'cultural image'. Section 5.4 then moved on to chronicle the ongoing spread of landscape phenomenologies through current cultural geographies, situating this within the broader discipline-wide turn to embodiment, practice and performance that has occurred under the aegis of non-representational theory. I noted in this section, and it is worth reiterating, that non-representational theory has only ever been a diverse and multiple set of arguments. Some of these arguments, in fact, run quite counter to phenomenological approaches, and in turn the final substantive section of this chapter (5.5) detailed some major continuing criticisms of the phenomenological approach. Moving forward in the light of these criticisms it can be argued that, on their own, the phenomenological insights of writers such as Ingold do not supply a wholly viable basis for landscape study by cultural geographers. In fact it might even be argued that these critiques of phenomenology, together with the anti-phenomenological currents of some work in non-representational theory, potentially pose something of a challenge as regards the continuing cogency of the landscape concept itself in cultural geography. Further discussion of this point, however, is reserved until the next chapter.

In conclusion, the argument can made that recent landscape phenomenologies, working out from the primacy of bodily practice and performance, have identified new topical grounds and new forms of research practice, at once enriching and diversifying the ambit of landscape studies within cultural geography in particular. I must declare a personal interest of course: my own writing thus far has sought to advance a broadly phenomenological approach to landscape. However, without risking too much objectivity, it is probably fair to say that landscape phenomenologies *have* opened up new avenues and agendas. This chapter has presented and discussed a rich array of examples of the new topics and grounds opened by landscape phenomenologies: stones, trees, ice, visions, orchard-growing, cycling, climbing, hiking, driving, gardening.

In presenting this extended discussion of landscape phenomenologies this chapter takes us close to the present day in terms of work on landscape in cultural geography. The past twenty or so years have witnessed not one but two revolutions in geographical landscape studies: first the inrush of successive waves of insight from visual theory, critical theory and poststructural thought (as discussed through Chapters 3 and 4), and second the advent of new phenomenologies of the body, materiality, perception and performance. The major task of this book has been to thoroughly detail these. In the next chapter, the book's final chapter, I want to look to the future of cultural geographies of landscape.

# 6

---

# PROSPECTS FOR LANDSCAPE

## 6.1 INTRODUCTION

This final chapter takes a prospective look at current work and emerging trends and agendas in cultural geographies of landscape. In terms of content, therefore, it has been the most difficult chapter to plan and write. This is because no one can know how cultural geographies of landscape will develop in the future – or even, it could be argued, if they will develop. It would have been very difficult to predict in 1980, for example, that in just over ten years a range of new interpretative approaches to landscape – landscape as way of seeing, landscape as text – would have galvanised work in cultural and historical geography in the UK in particular. Equally, from the vantage point of the early to mid-1990s, a resurgence of interest in phenomenological approaches to landscape would have seemed unlikely. Further complicating the task of assessing the present is the fact that today's trends do not necessarily constitute tomorrow's established research platforms. And beyond this, the fact that landscape has long been a key term and concept for geographers does not in and of itself guarantee its continuing purchase and salience in the future. But, as the preceding two chapters have demonstrated, landscape research is today very much a vibrant and evolving field and, given the diversity and scope of current landscape writing, one might quickly come to regret overly confident predictions about the future.

With all of this in mind, the tone here is necessarily somewhat circumspect. The aim of the chapter is simply to identify and document some of the main substantive foci of current work on landscape, and also to locate these in relation to cognate trends in human geography as a whole. Arguably, and interestingly, there is still a discernable geographical and cultural aspect here – that is, different understandings of landscape, and hence different substantive preoccupations, still predominate within different academic worlds, with (although this should not be overemphasised) a distinction between work in the UK and North America being especially notable. The chapter begins (Section 6.2) by commenting upon and attempting to account for this distinction, before moving on to identify a common thread in much contemporary North American landscape writing – one that emphasises the material and everyday qualities of landscape, and that focuses upon issues of memory, identity, conflict and justice. Following on directly from this, Section 6.3 discusses the relationship between landscape, life and law, a relationship brought to the fore in recent work by Kenneth Olwig in particular (Olwig, 2002, 2005a), and associated with what is in some ways a distinctively northern European and Scandinavian understanding of landscape.

Section 6.4 then returns to the forms of non-representational theory discussed in Chapter 5 in the context of landscape phenomenologies, but this time to make the argument that some of the trajectories being explored by geographers and others in this broad area are, in a way, *anti-landscape*, and potentially impact upon landscape's role as an organising trope for cultural geographical writing. In particular here, and as I shall explain in more detail, work on the hybrid geographies of culture–nature relations (see Whatmore, 2002, 2006), explicitly seeks to advance a relational, vitalist and *topological* vision of culture and nature, and space and society, in which notions of landscape would seem to have little purchase. Section 6.5, by contrast, aims to illustrate the continuing cogency of UK-based landscape writing in the wake of non-representational theory, by discussing some examples of work that explores landscape via ideas of affect, presence, biography and movement. This is the last substantive section of the chapter, and the book as a whole. A summary conclusion follows, focusing upon the creative tensions involved in making and studying landscape.

## 6.2 MEMORY, IDENTITY, CONFLICT AND JUSTICE

### 6.2.1 Landscape geography in the UK and North America

While of course drawing on writing from other disciplines, the clear focus of this book has been work by geographers based in the UK and North America. However, in seeking to identify trends and examine concepts and approaches, relatively little has been said regarding the differences *between* UK-based and North American geographers vis-à-vis landscape. This is mostly because it would be invidious to suggest that there *are* any stark or simple distinctions here. But on the whole it can be argued that North American geography has a longer and better-established tradition of placing issues of landscape at the heart of human geographical research. Thus, for example, UK-based landscape geographers today have a tendency to reference the work discussed in Chapter 3 of this book – the critical cultural studies of the 1970s and the new cultural geographies of the 1980s – as a sort of baseline for landscape studies. North American work, of course, equally cites and looks to this literature, but in addition it often *also* continues to cite and draw inspiration from other, often older traditions. This is in no way to suggest that North American landscape geography lags behind its UK counterpart, rather it is to note that older traditions continue to supply a context for rethinking notions of landscape, place and so on for North American writers. Thus, for example, Adams *et al.*'s (2001) edited collection, *Textures of Place: Exploring Humanist Geographies*, looks to both the humanist and new cultural traditions to provide a forward-looking consolidation. Equally, and more explicitly, Wilson and Groth's (2003) collection, *Everyday America: Cultural Landscape Studies after J.B. Jackson*, offers both a retrospective and prospective account of landscape as everyday material world, in the context of J.B. Jackson's legacy (see Chapter 2, Section 2.4). And beyond this, just a browse through the recent Annual Meeting programmes of the Association of American Geographers (see: http://www.aag.org), will show that the cultural landscape and cultural ecology traditions, ultimately rooted in the work of Carl Sauer and the Berkeley School, are alive and well.

Noting these recent publications, it is perhaps possible to argue that there are at present, as there have always been, some differences in aim and emphasis between landscape writing in the UK and North America. Sections 6.4 and 6.5 below will discuss work mostly by UK-based (though not necessarily British!) geographers. Here, the focus is upon the North American context, and what Schein (2003) defines as a 'post-empiricist' approach to landscape. In this regard, considering in particular the Wilson and Groth (2003) volume and other recent books and articles, it can be argued that current North American landscape writing often has two distinctive aspects. First, there is a strong, almost intuitive sense that landscape be thought of as the everyday material world in which people live, work, interact and debate with each other. Thus, the public spaces, the streets and squares and suburbs of American towns and cities are a recurrently investigated empirical setting (e.g. Schein, 1997; Leib, 2002; Dwyer, 2004; Duncan and Duncan, 2003; Staeheli and Mitchell, 2005). This understanding of landscape as the shared and disputed matter of daily life speaks in one way of the continuing influence of J.B. Jackson's 'vernacular' approach (see Chapter 2), in which everyday American landscapes are, in a way, made visible as legitimate objects of academic concern and worth. More widely, work in this vein has drawn upon critical studies of everyday life such as Bourdieu (1977) and De Certeau (1984), in which the role of systems of regulation and governance are stressed alongside recognition of the ongoing, practical making of cultural and material worlds. But at the same time these geographies are characterised by the heavy stress they often lay upon substantial issues of power, identity, inequality and conflict as central ingredients of everyday material cultural and public landscapes. The work of the landscape geographer thereby becomes part of a broader movement advocating social change and justice. Here, the ongoing Marxist and radical landscape analyses of Don Mitchell (see Chapter 4, Section 4.3) are important and, in general, contemporary landscape writing by North American geographers explicitly pursues the politics of landscape, often via a mixture of a critical focus upon domination and hegemony, and a poststructural stress upon discourse and identity (see Rose, 2002b). Very much in this vein, for example, Richard Schein (1997, p.662) states that

> Landscapes are always in the process of 'becoming,' no longer reified or concretized – inert and there – but continually under scrutiny, at once manipulable and manipulated, always subject to change, and everywhere implicated in the ongoing formulation of social life.

Thus, while still writing within 'a geographic tradition broadly defined, yet coherent through its focus upon the *cultural landscape* as a tangible, visible entity' (ibid., p.660), Schein argues that 'the cultural landscape serves to naturalize or concretize – to make *normal* – social relations as embodied in the various discourses and their combinations' (ibid., p.676). Elsewhere, in his contribution to *Everyday America*, Schein (2003) explicitly grafts together the Jacksonian everyday and the forms of contemporary landscape politics discussed in Chapters 3 and 4:

> Post-empiricist approaches to cultural landscapes have begun to take seriously the assumption behind J.B. Jackson's concern for the landscape as capable of providing for social change. The cultural landscape is not merely the result of human activity. It is both a material thing and a conceptual framing of the world. . . . In short, the landscape is not innocent. Its role in mediating social and cultural reproduction works through its ability to stand for something: norms, values, fears and so on.
>
> (Schein, 2003, pp.202–203)

## 6.2.2 Landscape and the politics of memory

This notion of landscape, stressing its role as both material *and* discursive mediator of cultural values, has been extensively worked through in a recent suite of writings on landscape and the politics of *memory*. In fact, as Mitchell (2003b, p.790) notes, memory has been 'perhaps the strongest focus of landscape research in the past few years – landscape as a concretization and maker of memory'. A series of critical questions connect cultural landscapes up with memorial and heritage politics – at the most basic level, what is remembered and why? Who decides what monuments and other commemorative objects are erected in landscape? In what contexts does the exact location of such memorials become important? How do such memorial landscapes then function as markers of particular identities and particular historical narratives – as markers of who owns and controls the landscape? Conversely, whose memory is erased from the landscape? And how do repressed memories or officially forgotten events nevertheless return to haunt contemporary landscapes?

These sorts of cultural and political questions animate Karen Till's (2005) in-depth analyses of the politics of memory in Berlin, and also Dydia DeLyser's (1999) study of Bodie, a Californian 'ghost town' reinvented as a heritage tourism destination. Here, for example, the entire ex-urban landscape, rather than particular sites, streets or monuments within it, is produced by, and reproduces, a particular hegemonic cultural memory of the nineteenth-century American West:

> The false-fronted facades and ramshackle miner's cabins of this gold-mining ghost town call forth images to visitors from movie Westerns: heroic images of American pioneers. And since ghost towns like Bodie have few or no surviving residents, it is largely through the landscape of ghost towns, and the artefacts that are part of that landscape, that this American essence is apprehended.
>
> (DeLyser, 1999, p.603)

The point drawn is that such landscapes function as memorial sites in which dominant cultural values are asserted and reproduced. In this instance, the very materiality and hence seeming *authenticity* of the ghost town works to render alternative versions of Western history invisible and almost unthinkable:

> In the ghost town of Bodie . . . images of the past are not inclusive. The narratives of the mythic West of Anglo-American virtues, and of progress verify a patriarchal, middle-class Anglo-American construction of American culture, values and morals. In other words, what the majority of Bodie's visitors and staff find in Bodie is a place that confirms the already-held beliefs.
>
> (DeLyser, 1999, p.624)

The study of such landscapes of conflicted memory, heritage and identity has of course proved fertile for cultural and historical geographers in North America and the UK, in the latter finding expression through the mid-1990s in a sequence of studies of iconic landscapes of national identity (monuments, war graves, etc.) (Heffernan, 1995; Johnson, 1995; Withers, 1996). More recently the advent of postcolonial perspectives has turned attention to questions of decolonisation, diaspora and belonging, thereby giving impetus to the excavation of arguably subtler and more complex instances in which landscape, memory and identity intertwine – for example in Tolia-Kelly's (2004) work on the landscape values of British Asians, and in Jazeel's (2005) study of the nationalisation of nature in Sri Lanka. While explicitly postcolonial readings are rarer in North America, studies of landscape, race and remembrance have become almost a genre in their own right (Schein, 2006; Dwyer, 2000, 2004; Leib, 2002; Alderman, 2000). Many of these books and articles focus upon debates over the public commemoration of notable individuals such as Martin Luther King (Alderman, 2000) or Arthur Ashe (Leib, 2002), or alternatively take as their topic ongoing controversies over contested episodes such as the Civil War (e.g. Dwyer, 2004). For Owen Dwyer (2004, p.425), for example, in an interpretation that recalls the definitions of landscape given by Schein above, monuments are

> to be conceived of as in the process of becoming instead of existing in a static, essentialized state. Rather than possessing a fixed, established meaning, monuments are momentarily realized in a nexus of social relations as the result of attempts to define the meaning of representations, which nevertheless remain open to dispute and change.

And a particular vision of landscape resides in this citation and these literatures as a whole: landscape as both material entity *and* symbolic meaning, as both persistent in form and changeable in meaning, and thus as a key site for conflicts over memory, identity and justice.

To conclude this subsection, while the politics of memory has been perhaps the standout theme of recent North American landscape geography, it is important to note that this also is part of a wider critical geography, one inherited in part from the radical and Marxist geographies of the 1970s and 1980s, a critical geography whose key motifs are struggles for social transformation and justice. Questions of justice take centre-stage with respect to landscape in George Henderson's (2003) contribution to *Everyday America* (Wilson and Groth, 2003). As Mitchell (2003b) notes, this volume in some ways illustrates the manner in which North American landscape geographies remain at times uneasily hinged between descriptive and empirical traditions and critical and discursive concerns. Schein's (2003) and Henderson's (2003) chapters stand out a little amongst the others basically aimed at commemorating and recollecting J.B. Jackson. But, in asking the question, 'Are J.B. Jackson's celebrations of the great, improvised, vernacular spaces of America enough?' (p.195), Henderson responds with a clarion call reminiscent of Don Mitchell:

What is needed is a concept of landscape that helps point the way to those interventions that can bring about much greater social justice. And what landscape study needs even more is a concept of landscape that will assist the development of the very idea of social justice. . . . [T]he study of landscape, that thing which so often evokes the plane on which normal, everyday life is lived – precisely because of the premium it places on the everyday – must stand up to the facts of a world in crisis, to the fact that the condition of everyday life is, for many people, the interruption or destruction of everyday life.

(Henderson, 2003, p.196)

Mitchell (2003b) himself takes justice as the central theme of his final report on landscape for the journal *Progress in Human Geography*, and from that point offers a prospectus for landscape studies grounded in anti-racism, Marxist theories of power, scale and uneven development, and an empirical, evidence-driven social history. The way in which this will be worked out is a question for the future, but the terms of Mitchell's vision are clear:

Landscape studies not only can, but now must, develop the intellectual tools necessary to be part not only of the intellectual reinforcements needed to combat . . . degeneration, but also of the political reinforcements. Landscape studies can no longer be only about just landscape. Landscape is too important to be allowed, any longer, to be the dreamwork – or the groundwork – of empire. Landscape studies must be dedicated to seeing that landscape becomes the groundwork – and dreamwork – of justice.

(Mitchell, 2003b, p.793)

## 6.3 LANDSCAPE, POLITY AND LAW

In the course of his plea to couple landscape and justice, Mitchell raises the associated question of the relationship between *landscape* and *law*. This question, he notes, has been a central theme of the work of the 'pan-Nordic' (Mitchell, 2003b) geographer Kenneth Olwig (1996, 2002, 2005a, b, c), and he goes on to describe a seminar in Trondheim, Norway, in which both he and Olwig had participated. Here, he writes, 'landscape was something quite different than what we had come to think of it as in Anglo-American geography' (Mitchell, 2003b, p.792). Olwig's work has already been briefly touched on (see Chapter 4, Section 4.2) in the context of new cultural geography's anxieties over the materiality of landscape. This section will be devoted to further discussing the distinctiveness of his own understanding of landscape in terms of polity, custom and law. A focus upon one individual's work is merited here, I think, because Olwig's writing represents perhaps the most ambitious recent attempt to thoroughly redefine the basic terms of landscape, its origins and its ambit as a conceptual tool. In turn, as will be discussed below, this is also now inspiring new substantive work on landscape and law (Olwig, 2005a).

Olwig's redefinition of landscape, Mels (2003, p.381) notes, is based in an attempt to recover 'the now ghostly and extremely complex traces of the original "Germanic" definition of landscape as a tract of land, regarded as a territorial and customary unit'. As this implies, Olwig (2002) attempts to work out, etymologically and discursively, an understanding of landscape that is in many ways distinctive from all those examined through the course of this book. Landscape here is not an empirical field to be studied morphologically (Chapter 2), nor is it a way of seeing the world (Chapter 3), or an associable set of cultural values, practices and governances (Chapter 4), or an ensemble of dwelling practices (Chapter 5). Instead, via an examination of the German word *Ländschaft*, the root of the present-day English word landscape, Olwig in a sense attempts to look *behind* and *before* all of these formulae, to recover an original meaning for landscape, one which can then inform contemporary critical thinking. Originally, Olwig argues, long before the invention of perspectival painting led to its pictorial and scenic redefinition, landscape was *a particular sort of legal and political entity*: 'the primary meaning of *Ländschaft* appears to have been a judicially defined polity, not a spatially defined area' (Olwig, 2002, p.19). To put this another way,

> custom and culture defined a *Land*, not physical geographic characteristics – it was a social entity that found physical expression in the area under its law. . . . The *Land* was [thus] initially defined by a given body of customary law that would have developed historically within and through the workings of the judicial bodies of a given legally defined community.
>
> (Olwig, 2002, p.17)

This claim on behalf of landscape has a particular geographical and historical context – that of pre-Renaissance northern Europe – and Olwig's most comprehensive statement thus far, *Landscape, Nature and the Body Politic* (2002), is, like Cosgrove's (1998) *Social Formation and Symbolic Landscape*, a historical account of how the meaning of landscape has changed over time and through space. In its original context, Olwig argues, landscape was understood above all as referring to a political community of people – a polity – and the set of customary, local laws through which they administered themselves. The term landscape thus denotes 'a nexus of law and cultural identity' (Olwig, 2002, p. 19). Northern Europe as a whole was once composed, Olwig goes on to suggest, of myriad diverse local polities, or landscapes (*Ländschaft*), some larger, some smaller, although all were increasingly coming under the judicial and sovereign sway of centralising monarchies and nascent nation-states. The overall narrative of *Landscape, Nature and the Body Politic* is thus one in which landscape as local polity and place is gradually supplanted from the sixteenth through to the nineteenth century by a scenic, pictorial and formal definition – a definition which corresponds with the official discourse and legal tenure of modern nation-states. Landscape moves 'from common*places* to scenic *spaces*' (*ibid.*, p. 214, original emphasis). The point, therefore, is not to deny that landscape has been, and is, an ideological way of seeing, but rather to offer a corrective to cultural geographies (e.g. Cosgrove, 1985; Cosgrove and Daniels, 1988) which have assumed that to be the *only* genealogy of the concept. Equally, landscape conceived as polity and place has subsisted through the centuries, and today retains the ability to undercut and query dominant ways of seeing.

The wider point to draw, perhaps, is that defining landscape historically in terms of locality, community, polity and law is one way of granting the term a distinctive and enlarged purpose in the present day. Thus, for Mitchell (2003b, p. 788):

> What *Landscape, nature, and the body politic* provides . . . is, along with an important historical excavation of the roots of our contemporary landscape way of seeing, an equally important excavation of a quite different way of seeing landscape's relationship to law and social justice. [Olwig] suggests that beneath the dreamwork and groundwork of empire lies a very different relationship between people and their landscape, one that is never fully repressed: there is a struggle for landscape, and it is at the same time the struggle for justice.

A story of landscape told as a vernacular 'people's' history of cultural use, value and transformation is certainly one message of Olwig's work. But, as Mels (2003) notes, this work emerges at least as much from the traditions of humanistic geography and historical landscape scholarship as from Marxist thought (and it is worth noting that Olwig chose Yi-Fu Tuan, a luminary of geographic humanism, to write the foreword to *Landscape, Nature and the Body Politic*). In other words, the ethos here is anti-structuralist and anti-abstraction – what Olwig pursues is a history made by people rather than abstract forces. Thus, 'the landscape/country as a physical place was . . . the manifestation of the polity's local custom and common law' (2002, p.214). And further in this vein, to capture a sense of landscape as an evolving set of human cultural practices and customs, Olwig refers to Pierre Bourdieu's (1977) notion of habitus, in stating that 'landscape is the expression of the practices of habitation through which the habitus of place is generated and laid down as custom and law upon the physical fabric of the land' (2002, p.226). This stress upon law and custom, it should be noted, sharply differentiates this account of landscape from what Olwig calls 'blood-and-soil nationalism': landscape is a historical and cultural entity, made through law, not nature – it belongs to a polity, not a species.

A certain humanist ethos is also evident in Olwig's sense of how his vision of landscape might play out in present-day contexts. Rather than a politics of radical transformation, he speaks of humanised philosophies informing better planning and design policies:

> Architects who think only in terms of the power of scenic space, ignoring the exigencies of community and place, run the risk of producing landscapes of social inequality like those of the great eighteenth-century British estates. . . . It is also possible, however, for architects to shape environments that foster the desire to maintain the continuities that maintain a collective sense of commonwealth, rooted in custom but open to change.
>
> (Olwig, 2002, pp.226–227)

This turn to questions of policy, planning and design further informs and guides a special issue of the journal *Landscape Research* on landscape and law, guest edited by Olwig (2005a). Here he writes that 'emphasis is now shifting from a definition of landscape as scenery to a notion of landscape as polity and place' (p.293). This perhaps pushes the case a little, and risks being seen as an attempt to effect a change simply by announcing it, but Olwig further points to the definitions worked through in the Council of Europe's European Landscape Convention (2000), and argues that these supply a distinctively cultural, political and legal (as opposed to physical or environmental) basis for approaching landscape. The special issue itself emerges from the seminar discussions in Trondheim mentioned at the start of this section, and rests upon 'a common discursive ground in discussions of the meaning of law and justice in relation to landscape' (ibid., p.295). What is perhaps most notable here, in terms of prospects for landscape studies going forward, is the mixture of 'discursive' and 'applied' landscape writing that devolves from this common ground. Olwig's (2005b) own contribution is a more meditative essay on landscape and customary and natural law, but alongside this there are accounts of how landscape-as-polity informs contemporary planning and administration in Holland (Mels, 2005), engagements with conflicts in public landscapes over access, community and justice (Staeheli and Mitchell, 2005; Spirn 2005), and studies of legal practice in relation to landscape matters (Martin and Scherr, 2005). And the thread linking these diverse topics is clear: 'law plays a central role in the constitution of landscape' (Olwig, 2005a, p.296).

## 6.4 THE ENDS OF LANDSCAPE? RELATIONALITY, VITALISM AND TOPOLOGICAL GEOGRAPHIES

In focusing upon issues of memory, justice and law, much recent work by North American and northern European geographers has a substantive feel – it has concentrated, for the most part, upon 'grounded' studies, rather than elaborating further concepts of landscape, and it is concerned with the affirmation of interpretative and discursive arguments regarding landscape within various tangible and physical contexts. As Mitchell (2003b) notes, this empirical work follows a period up until the late 1990s in which new theoretical and conceptual languages for landscape were extensively elaborated. By contrast, however, with the advent of various non-representational approaches (Chapter 5, Section, 5.4), work in UK-based human geography since the later 1990s has been characterised by continuing and even accelerating theoretical developments. In this section, therefore, I want to follow up on some of the themes and issues discussed in the previous chapter in the context of landscape phenomenologies. It was noted there that some of the avenues being explored by cultural geographers, in particular as regards culture–nature relations, could be viewed as in some ways antithetical to the projects that have concerned geographical landscape studies – antithetical, that is, to both the interpretative approach emphasising landscape's role within systems of cultural discourse and practice *and* the phenomenological approach in which lived experience is foregrounded. This section will flesh out that argument. In doing so it will tackle ideas and terms which are arguably quite complex, but which are also, it must be stressed, increasingly used and discussed in many branches of human geography. In other words, this section does not deal with ideas that are esoteric or speculative, instead it aims to tie together some of the actual themes and topics that cultural geographers, especially in the UK, *are* currently writing about. First of all, then, it will outline arguments concerning the inherently *relational* nature of identities, geographies, economies and natures themselves. This relational argument has been widely taken up across human geography since the late 1990s, often accompanied by a parallel stress upon process, movement and becoming. The vision here, as mentioned in the previous chapter, is of geographies continually on the move and the making, connected and connecting. This relational and connective vision further owes much to the philosophy of *vitalism*, and the section will second briefly locate this idea in the work of the philosopher Gilles Deleuze. Moving on, I will discuss emergent 'hybrid geographies' (Whatmore, 2002) of nature and culture, in which relational and vitalist principles are brought together. And, via discussion of this still-emergent literature, the argument will be made that such hybrid geographies paint a *topological* picture of the world in which received notions of landscape would seem to have little place.

## 6.4.1 Relationality

First, then, relationality. Much contemporary human geography, in stressing the significance of networks, connections, flows and mobilities in the ongoing making of spaces, places and identities, could be said to subscribe to a *relational* vision of the world. There is here a general argument for 'thinking space relationally', such that it is viewed as 'a product of practices, trajectories, interrelations' (Massey, 2004, p.5). And this sense of space as a weaving and a relating, forever in the making, sets out to critique and supplant arguably more static notions of space in terms of territory, boundedness, area, scale and so on (Marston *et al.*, 2005). Notions of relationality are thus being worked out in a variety of empirical contexts, for example in a newly relational vision of economic development (Yeung, 2005), in work on commodity chains and material cultures of production/ consumption (Hughes and Reimer, 2004), on transnationalism and cities (Olson and Silvey, 2006), and on the possibility of a 'new mobilities paradigm' (Sheller and Urry, 2006).

The most important thing to grasp, perhaps, is the manner in which relations are being viewed as primary in these arguments, as being that which fundamentally exists. Rather than relations and connections being forged in an already-given space, relations are being viewed as *creative of spaces* (as the quote above from Massey implies). That is, relations do not occur in space, they *make* spaces – relational spaces, and the geography of the world is comprised of these. This relational view of a world-always-in-the-making owes much of its derivation to a set of arguments called actor-network theory. Actor-network theory – or, as it is invariably abbreviated to nowadays, ANT – is a coalition of propositions which originally emerged from ethnographic studies of science and technology (e.g. Latour, 1987). Two of the earliest works most often cited as informing the agenda of ANT – Bruno Latour's (1993) *We Have Never Been Modern* and Donna Haraway's (1991) *Simians, Cyborgs and Women* – were concerned above all to disturb and unsettle the academic (and also commonplace) habit of dividing the world up into 'cultural' (or social) and 'natural' domains. Both Latour and Haraway argue that this is essentially an attempt to separate off a purified space of the human and the thoughtful, untainted by materiality or animality. Its consequences, however, are manifold; for example we assume the primacy of individual, discrete subjects, and we locate agency, will, creativity and the capacity for action solely *within* the human subject. One of the central tasks of ANT has been to query these assumptions and to argue that the notion of two separate realms, natural and cultural, passive and active, is no more than a beguiling fiction, albeit one that has had profound consequences.

First, then, ANT is concerned with querying the human/non-human divide. It argues that this divide is untenable and that there are no such things as purely 'natural' or purely 'cultural' entities. Instead we are all of us, already, *hybrids*, complex amalgams of human/animal/machine (Haraway, 1991). As will be discussed below, this has been very important in the development of 'hybrid geographies'. Once recognition of hybridity has been made, ANT moves on to argue that we should take non-human entities – plants, animals, oceans, aircraft – very seriously, according them no less than equal status to human beings. Non-human entities, in this account, have as much ability to act and be influential as humans.

However, while ANT wants to take non-human objects and processes seriously, it also wishes to query our tendency to conceptualise the world in a monadic fashion, in other words to view the world as being composed of discrete, bounded entities. We have long mistakenly thought, ANT suggests, in terms of discrete subjects and objects, people and things. In contrast to this conception, ANT, as the name implies, foregrounds *networks*. The world as pictured by ANT is composed first and primarily of networks, relations and connections and only secondarily and consequentially of discrete objects and actors or 'actants'. Every object, item or individual can be understood as the assembled outcome of networked relations; indeed our sense of our individuality and agency is but a relational effect. Instead of being in-dividual, that is, in-divisible and whole, we are always already 'dividual' – we are woven together through and amidst a complex *mélange* of at once cultural and material, human and non-human relations. In other words, what ANT proposes is a *relational ontology*.

## 6.4.2 Vitalism

If notions of relationality derived in part from ANT are currently high on the agenda of human geography, another, associated set of arguments emerges from the philosophical tradition of *vitalism*, and, in particular, the work of Gilles Deleuze. Deleuze's varied writings on philosophical history, art, cinema and psychoanalysis emphasise creativity and transformation, not just as values to be cherished, but as immanent, ontological features of life in general. In this sense, Deleuze's philosophy, and in particular his collaborative work with the psychiatrist Felix Guattari (1988), has often been characterised as 'philosophy of becoming' or as a 'new vitalism' (Marks, 1998). The notion of *becoming* first captures the Deleuzian sense of a world continually in the making, continually proliferating. It also captures the strongly anti-phenomenological bent of Deleuze's writing; in so far as 'becoming' is explicitly a radical alternative to what Deleuze would see as the static and sedentary tonalities of Heideggerian notions of dwelling and 'being-in-the-world'. And, lastly, the Deleuzian notion of life as a blossoming and differentiating becoming (becoming-animal, becoming-woman, becoming-imperceptible), at the very least rhetorically chimes with ANT's vision of a world of hybrid relationalities.

In turn, the term 'vitalism' captures a particular ethos – a sense of life in general – as an immanent, vital, emergent force. Again, to refer to the work discussed in the previous chapter, this vitalist ontology in many ways countermands phenomenological perspectives, through advocating a radically non-subjective account of life, one in which the circulation of vital non-personal and non-human affects, forces and singularities is understood to supersede, rather than supplement, notions of dwelling and being-in-the-world. What is envisaged is an understanding of *matter* – the materiality of things and of nature – as alive and animate. The world Deleuze pictures is 'a pure flow of life and perception, without any distinct perceivers' (Colebrook, 2002, p.87). And in this way, Deleuze's account of perception and feeling aims to go 'beyond the phenomenological anchoring of a sub-ject in a landscape' (Rajchman, 1998, p.10). So, in sum, Deleuze's writing supplies at the very least a strong corrective to any landscape phenomenol-ogy reliant upon unexamined notions of subjectivity, perception and knowledge. It might even be considered to invalidate and supersede such a phenomenology.

### 6.4.3 Topology and hybrid geographies

One important argument for contemporary hybrid and biogeographies – the field of enquiry where the relationalities of ANT and the vitalities of Deleuzian thought now most productively coalesce – is that we are increasingly living in a post-humanist world. New biotechnologies and imaginaries, the manipulation at the genetic level of humans, animals and other organisms, blur boundaries long held secure and even sacred. Sarah Whatmore's (2002, p.1) *Hybrid Geographies*, a leading text in this area, thus opens with remarks upon 'the hyperbolic inventiveness of the life sciences to complicate the distinctions between human and non-human; social and material; subjects and objects'. She goes on to argue that, partly in order to comprehend and adequately analyse this emerging reality, a turn towards post-humanist modes of enquiry, is necessary. The production of hybrid geographies will perforce involve: 'shifts from intentional to affective modalities of association; from being to becoming in the temporal rhythms of human/non-human existence; and from geometries to topologies as the spatial register of distributed agency' (Whatmore, 2002, p.5).

Most recently, in a survey of 'materialist returns' in cultural geography, Whatmore (2006, p.602) argues that 'a new generation of cultural geographers is returning to the rich conjunction of the *geo* and the *bio*', in a new biogeography of life itself, positioned at the heart of geography (human and physical). This subsection will not examine in detail the work by Whatmore and others writing in this vein about issues raised by new biotechnological developments (e.g. Greenhough and Roe, 2006; Clark, 2002; Hinchcliffe *et al.*, 2005; Thrift, 2006; Bingham, 2006). Instead I want to examine the potential consequences this seemingly emerging paradigm might have for notions of *landscape* specifically. First of all, as should hopefully be clear from the discussion of relationality and vitalism above, work in this vein is, like the landscape phenomenologies discussed in the previous chapter, very strongly opposed to any notion of nature as a 'cultural construct', and more generally opposed to the discursive and interpretative approaches characteristic of much cultural geography through the 1990s. Hybrid geographies therefore reject the idea that landscape is a way of seeing, a gaze projecting cultural meaning onto an inert material nature. As Whatmore (2002, p.1) argues, in this conception, 'nature, having nothing to say for itself, is the always already crafted product of human interpretation'. Elsewhere, she notes that both 'old' and 'new' cultural geographies 'cast the making of landscapes (whether worked or represented) as an exclusively human achievement in which the stuff of the world is so much putty in our hands' (Whatmore, 2006, p.603). But it can also be argued that hybrid geographies share equally little common ground with landscape phenomenologies. This argument can perhaps best be sketched out through reference to *topology* – the movement 'from geometries to topologies as . . . spatial register' described by Whatmore above.

Topology is the study of object properties which are maintained even when the object is stretched, compressed, twisted or otherwise altered (but not torn). More loosely, topology is a form of geographical/mathematical thinking that conceives space and spatial relations primarily in terms of connective properties rather than distance and position. For example, the London Underground map is topological. It shows how tube stations are connected and interrelated but ignores as irrelevant to the user the real distances between stations and the actually curving pathways of the trainlines.

Theorists of space and spatiality writing from an ANT perspective have commonly argued for a topological understanding of geography (Law, 1999). ANT *in toto*, it is argued, 'concerns itself with the topological textures which arise as relations configure spaces and times' (Murdoch, 1998, p.359). And in some ways this is simply another way of saying that relations come before positions. Instead of thinking of the surface of the earth as a holistic bounded area (the topographic perspective which underwrites both cartography and our everyday assumptions about how space exists) in which connections are made, we need, the argument runs, a new geographical imagination in which topological connections are the primary 'weavers' of space–time. Topology thus implies that space and time are not geometric and linear. For example, given the volume and density of connections between the two cities, London is closer, topologically, to New York, than it is to other towns and cities in the UK.

A topological and connectivist ontology is clearly at work within today's hybrid geographies. The vitalist ethos informing this movement further implies not simply a fixed or already-given set of connections, but rather a burgeoning, proliferating, even wondrous topology, in which attendance to uncanny and hybrid foldings of near and far and past and present becomes the crucial critical task. For Whatmore (2002, p.6), therefore:

> In place of the geometric habits that reiterate the world as a single grid-like surface . . . hybrid mappings are necessarily topological, emphasising the multiplicity of space-times generated in/by the movements and rhythms of heterogeneous association.

The point to draw here is that such a topological geography exists in a state of tension vis-à-vis concepts of landscape, for several reasons. First and most obviously, in contrast to the fluidity, mobility and heterogeneity prized in the topological paradigm, notions of landscape can seem static and 'sedentarist' (Sheller and Urry, 2006). In general, with its discursive associations with dwelling, remaining and envisioning, landscape folds uneasily with movement, even though, as the previous chapter discussed, the making of landscapes and selves through reciprocal movements has been a key issue for recent writing (e.g. Cresswell, 2003; Wylie, 2005; Büscher, 2006; Lorimer, 2006). Second there is the issue of succession in so far as new biogeographies have to some extent supplanted landscape as a medium for thinking through culture–nature relations, in particular superseding and critiquing notions of landscape as a cultural 'construction' of nature (Whatmore, 2006). Thirdly and lastly, landscape has of course historically been very much an areal and *topographical* term, – the shape of the earth's surface – and has long been affiliated with precisely the conceptions of space, measure, distance, surface and perspective that topological geographies want to make strange.

In sum it seems difficult to accommodate landscape, with all its topographical, visual, phenomenal and synthetic associations, within a topological vision. Certainly it is the case that some proponents of the new topological geography seem intent upon discarding 'middle terms' or synthetics such as landscape. Whatmore (2006), for example, explicitly locates material and hybrid geographies at some distance from the cultural geographies of landscape advocated by figures such as Denis Cosgrove and Stephen Daniels. Another recent hybrid text *Patterned Ground: Entanglements of Nature and Culture* (Harrison *et al.*, 2004) also aims to move clearly in this direction. This text eschews all conventional notions of order and classification. It begins, for example, with the sequence, 'Pipes, Cities, Ecosystems, Rivers, Ridges, Cliffs, Scree, Virtual Spaces, Post Offices, Drumlins'. And in taking such an approach, *Patterned Ground* is an original and inventive collection, offering over a hundred brief and often very lucid and expressive essays on a deliberately eclectic tangle of topics, written by authors from human and physical geography, social and natural science. Commenting on the sequencing of these essays, the editors note that they deployed a 'topological imagination' (2004, p.10) to purposefully unsettle and differently entangle conventional orderings of human and non-human landforms. And more broadly and significantly the argument here is that, far from being an image or construction of culture–nature relations, landscape *is* their very entanglement:

> We are arguing that there is a 'new geography' because it is important to appreciate that the world is now patterned by both human and nonhuman processes. It is to these entanglements – that comprise what we know as landscape – that this book is oriented.
>
> (Pile *et al.*, 2004, pp.9–10)

In many ways, it should be noted, this is a project that the landscape phenomenologists discussed in Chapter 5 might well endorse. It recaptures, for example, a sense of the dynamic materiality of landscape. It also thwarts constructivism and discursive idealism. But it can also be argued that, in *Patterned Ground* and other hybrid geographies, a certain topographical richness is being sacrificed for the sake of topological complexity. In other words, in prioritising relations and trajectories, such topological and vitalist geographies present what is a curiously flat and depthless picture. The ground might be patterned, but it is *flat*. It ironically resembles a cross-hatched isotropic plain – a Christaller space. Here every point, every object, is accorded an equal weight and value (other essay sequences in *Patterned Ground* include: 'Bees, Pubs, Pigs, Humans, Moon' . . . and 'Jungles, Slums, Buildings, Archives, Streets'). All equally cede to the primacy of the relational and the connective. And the result, it can be argued, is a sort of ontological over-flattening. To put this another way, while hybrid geographies have undoubtedly opened up new and rich ways of thinking about life, subjectivity, perception, culture and nature, we are perhaps being presented with a topology without topography – a surface without relief, contour, morphology or depth.

## 6.5 LANDSCAPE WRITING: BIOGRAPHY, MOVEMENT, PRESENCE AND AFFECT

### 6.5.1 Landscape writing

If, as the previous section suggested, some recent trends are seemingly working against the continuing salience of landscape as both organising trope and source of creative and productive tension for cultural geographers, then it is equally possible to argue the opposite: in the early years of the twenty-first century studies of landscape are burgeoning, and current work by UK-based geographers is developing new and distinctive approaches to landscape. For example, Chapter 5, Section 5.4 discussed a wide range of recent work in which landscape is reworked via performative and embodied conceptions of tactility, mobility and visuality. This section will look more closely at work which is affiliated to such concerns, but which is further characterised by a focus upon *landscape writing*. In the wake of insights from poststructural and non-representational theories, it may be argued there is a renewed sense of a need to develop newly critical and creative means of expressing relationships between biography, history, culture and landscape.

Of course, at various points in this book, issues of writing and the self have already loomed large in relation to landscape. The work of early landscape scholars such as Sauer, Hoskins and Jackson has endured, it could be argued, in part because of the distinctively personal quality of their landscape writing, its authority and rhetorical power. Chapter 3,

Section 3.5 further discussed approaches in which landscape is more generally conceived as a text, as something written and read. Chapter 4, Section 4.5 also worked through critical literatures in which the verbal and visual depiction of landscape in the context of European travel writing formed a central object of analysis. And Chapter 5, Section 5.4 detailed the varied ways in which embodied practice and performance has been understood as a milieu through which entwined senses of self and landscape emerge.

More broadly, beyond the confines of academic studies, it could be argued that notions of landscape – the description of landscape, and the discussion of personal experiences of and attachments to landscape – are generally consonant with the literary genres of nature writing, travel writing, cultural history and biography. And in this context, the last ten to fifteen years have been something of a golden age for landscape writing at large, with the publication, for example (and this list could doubtless be extended), of Tim Robinson's (1990, 1995) monumental accounts of the landscape and history of the Aran Islands off the west coast of Ireland, Rebecca Solnit's (2001) polemical accounts of landscape, walking and the self, Iain Sinclair's (1997, 2003) varied excursive wanderings in and around the landscape of London and England.

The work of the author W.G. Sebald (1994, 1998, 1999, 2001) perhaps merits particular mention here, in so far as it has been a specific source of inspiration for recent geographical engagements with landscape writing (e.g. Dubow, 2000, 2004; Lorimer, 2003a, b, 2006; Wylie, 2002a, 2005). German-born, but writing also from an academic position as a Professor of European Literature, and working until the late 1990s at a UK university, Sebald's writing, especially his existential walking tour through and beyond East Anglia, *The Rings of Saturn* (1998), has come to stand as something of a model for contemporary cultural geographies of landscape. Substantively Sebald's books insistently return to, and circle round, a number of linked themes: exile and displacement, memory and forgetting, war and violence, solitude and connection, and history, both human and natural, considered as a process of destruction and fragmentation. But Sebald is perhaps most notable and influential for his elaboration of an innovative literary form, one incorporating elements of existential memoir, autobiography, travel writing, cultural history and phantasmagoria. In and through this format Sebald conjures a strange metaphysics of landscape, one that succeeds in both pressing together *and* unravelling past and present, text and image, experience and memory. Like landscape itself, therefore, we are presented with fact, fiction, travelogue, memoir and lament, all at once.

A large measure of Sebald's originality further lies in his attentiveness to the tension between movement and rootedness that has always haunted landscape studies. As has been noted at several points through this book, many cultural geographers have viewed landscape with suspicion because of the term's seemingly inherent conservativism, its association with dwelling and remaining, with closed and local horizons, as opposed to a more cosmopolitan or mobile perspective. In turn, however, it could be said that landscape is always only ever a question of movement, movements of eye and body, land and sky, movements inwards and outwards, if only these could be faithfully described. Drawing upon a European critical and avant-garde legacy, including manqué writers such as Maurice Blanchot, Elias Canetti and, especially, Walter Benjamin, W.G. Sebald provides what is a quite original treatment of landscape's essential tension in-between movement and dwelling, outsider and insider. What is most beguiling here is the precision of this tension. Caught precisely between staying and moving on, Sebald's writing is *of* landscape, not simply *about* landscape, and so still remains committed to the critical and perspectival possibilities vouchsafed by what Jessica Dubow (2004) calls 'the mobility of thought'.

## 6.5.2 Biography and movement

This co-scripting of landscape, movement and biography has become a key theme for recent landscape writing. In a series of papers, for example, Hayden Lorimer (2003a, b, 2006) seeks to creatively and critically address the ways in which land, life and knowledge are entwined. Lorimer is especially concerned to work through, and re-present, an original nexus of personal memories (accessed through oral and archival history), intellectual histories of the field sciences, and embodied experiences and evocations of landscape. What is sought here is a new form of geographical historiography. And the agenda underpinning this approach is very much driven by a perceived need for cultural and historical geographies to develop new critical styles, in the wake of non-representational theory's critique of the 'deadening' effect of a thematic and discursive focus upon texts and images:

> The key requirement is a creative engagement with, and imaginative interpretation of, conventional 'representational' sources, rather than

the identification of a previously ignored or oppositional realm of non-representational practice.

(Lorimer, 2003a, p.203)

Thus, the paper from which this quote is taken is very much a creative work of historical reconstruction. Aiming to 'tell small stories', as a counterpoint to the grander narratives of academic history, it stitches together the life and work of Robin Murray, a field instructor and former geography student, working at a lodge in the Cairngorm Mountains in Scotland, with the recalled experiences of an elderly woman, Margaret Jack, who as a 15-year-old Glaswegian city girl had participated in an outdoor fieldtrip led by Murray. Here, therefore, poignant personal memoir enlaces productively with the bodily experience of the Cairngorms, and with the knowledge practices of 1950s geography, then still rooted in survey and classification. Elsewhere, in 'Herding memories of humans and animals', a piece more specifically focused upon landscape, Lorimer (2006) recounts another episode in the recent history of the Scottish highlands – the relocation of a herd of reindeer from Finnish Lapland to the Cairngorms – and in doing so moves a little further still from the standard protocols of historical scholarship:

> You have not studied with your own eyes the long, upward sweep of the land that lifts sheer at the northern corries and then once on high stretches out across the granite expanse of the mountain plateau, nor the stands of pines that survive on the lower slops, nor the sharply incised ravine that must be crossed to reach the grazing grounds. And, since I cannot take for granted that you know this topography and its particular brand of local information, these responses require careful animation.
>
> (Lorimer, 2006, pp.497–498)

In approaching this topography, and the lives of both the reindeer herd and those humans involved with it, Lorimer thus aims to at once ask and answer 'searching questions: how best to encounter the textures and cycles of work that leave a landscape replete with meaning? What creative strategies might be employed to reanimate, however temporarily, the embodied relationship between individual subjects and an environment?' (ibid., p.504).

The answer, in this case, is the device of a love story. While much of 'Herding memories' is devoted to exploring the bodily experience of the herd and herders upon the Cairngorm plateau – an experience that 'takes shape in flanks, lips, shoulders, rumps, spurs, necks, dips, pools, cuts and brows' (ibid., p.505) – the heart of the paper lies in the tale of the relationship between a Cambridge anthropologist, Ethel John, and a nomadic Lapp herder, Mikel Utsi. This relationship was central to the transportation of reindeer to Scotland but, perhaps more crucially, through this device, Lorimer is able to express landscape as a matter of movements, of biographies, attachments and exiles. The narrative weaves between Cambridge, Scandinavia and the Cairngorms, and while landscape is thus lifted from locale and identity in one sense, in another it is creatively re-woven in the forms of conflicting affections and passages of movement-between. Local knowledges and the sensuous involvement of the herd and herders with grass, water, rock and sky, are so enriched by a transnational tale of distance and intimacy. Thus Lorimer argues for a form of landscape study at once empirical and historical and affective:

> The footwork and field trudge may remain the same but the manner in which landscape is approached and expressed can be retuned to shifting, sentient encounters, to fuse material and mental landscapes, and to telescope down to share in the spatiality of individual lives lived.
>
> (Lorimer, 2006, p.516)

And this focus upon 'lives lived', upon biography and narrative, becomes a leitmotif for landscape writing:

> Remaining faithful to those registers of memory where a past landscape takes shape as oral tradition, as embodied knowledge, or through shared personhood, radically reframes the challenge of reconstruction and presentation. Here, it is the possibility of crafting a closeness to the style and tone in which events are remembered, located, and organised that becomes of greatest moment. Landscapes told as a distribution of stories and dramatic episodes, or as repertoires of lived practice, can be creatively recut, embroidered, and still sustain original narratological integrity.
>
> (Lorimer, 2006, p.515)

For Lorimer landscape writing is at once a mossy, earthy endeavour, *and also* a form of critical historical practice, in which shuttling movement between different spaces and registers outflanks a tendency to naively valorise rootedness. This notion of landscape and self as essentially written through experiences of mobility and exile also features strongly in the work of Jessica Dubow (2000, 2004). Here, however, in partial contrast to Lorimer's poetics, the aim is to supplement a phenomenological stress upon experience and encounter with insights from Continental critical theory. Like Sebald, Dubow looks to figures such as Maurice Blanchot and Walter Benjamin – representatives of a diasporic and Judaic philosophical tradition – for inspiration. And again like Sebald, the aim here, in an account of identity and knowledge as *originally* formed via mobility, exile, distance and non-belonging, is to precisely *displace* notions of landscape and self that are based upon the defined presence of a perceiving subject and the identification of that subject with a native soil or nation-state.

For Dubow, a critical landscape writing requires a centripetal, outward-looking movement, or, more exactly, an original mobility that is wholly exterior and anterior to what she regards as the territorial and bounded binaries of Western thinking – inside/outside, self/other, dwelling/nomadic – and even to the teleological notion of 'origin' itself. Importantly here, the gathering of a certain Judaic lineage of thought gestures towards a critical politics of distance and dislocation that is quite distinct from the vitalist ethos of becoming discussed in the previous section:

> what I propose here is not any utopianism which privileges perpetual 'becoming' as the ideal form of subjective practice; a position which, in the manner of certain interpretations of Deleuze and Guattari, rejects the territorial in its haste to equate nomadic mobility with the powers of a subversive critical capacity.
>
> (Dubow, 2004, p.219)

### 6.5.3 Presence and affect

As with Dubow and Lorimer, the work of Mitch Rose (2002b, 2004, 2006) – another cultural geographer concerned with the question of landscape writing in the wake of non-representational theories – also explores the interrelations of landscape and movement. Here, however, movement is not figured as a displacing of notions of origin, territory or dwelling, nor as a criss-crossing of life histories and landscape affinities. Instead the direction of travel is, so to speak, inward: landscape, for Rose, is conceptualised in terms of an enfolding and creative movement of care. And this conception stems in part from a critique of the forms of contemporary North American landscape analysis discussed in Section 6.2 above. For Rose (2002b, p.459), despite their ostensible embrace of a poststructural understanding of power and identity, these geographies remain fundamentally structuralist:

> On the one hand the landscape is a cultural symbol that can be diversely interpreted, and on the other it is a stable image whose existence depends on its interpretation being contained. Although struggles in space affect, disrupt and even re-write the hegemonic ideologies that produce the landscape, they do not in themselves define the landscape. . . . Thus, while landscape is described in terms of struggle, it is defined in terms of structure.

At the same time, much North American (and indeed UK-based) cultural geography emphasises landscape as a terrain of *contested* meaning, and this, Rose notes, relies upon the ability of actors to resist, ignore or disrupt the imposition of dominant, already-structured meanings. Here, therefore, there is a dilemma, a contradiction even: 'one cannot derive highly interpretative social agents from a society that is fundamentally structured. Neither can one derive fundamental structures from a highly interpretative society' (ibid., p.460).

Rose (2004, 2006) seeks to move beyond this problematic blend of structuralist and humanist positions by exploring instead the ways in which enduring and imaginatively compelling ideas of landscape and self are cultivated. In particular, to advance upon a notion of culture as a code to be broken, he argues, we must learn to 'recognise not only the movement of deconstruction but also the movement of what Derrida calls our "dreams of presence": our dreams of being a subject' (Rose, 2004, p.465). Landscape and self, for instance, may be understood as 'dreams of presence' – not as actual, stable or pre-given presences, but as nevertheless constituent parts of an incessant, nurturing or caring movement-process, in which the world is imagined as whole and coherent. For Rose (2006, p.547):

> Conceptualising the cultural landscape as a dream of presence means understanding it as an unfolding plane of sensory, affective or perceptual markers registering and, thus, effecting the emergence

of subjectivity. Yet, critically, it is also an *active* depositing of those markers through the movement of care.

This 'gathering' movement of care, Rose suggests, may be witnessed in narratives and biographies, and stress is therefore laid here upon creative and imaginative acts of 'storytelling' – of narrating and representing – via which landscape and self are dreamt. A study of West Virginian storytelling by the ethnographer Kathleen Stewart is presented as exemplary here, in so far as it 'is a form of writing that although not being fiction, in the sense that it is not self-consciously inventive, is nonetheless imaginative' (*ibid.*, p.537). In an echo of Lorimer's arguments above, Rose thus calls in conclusion for a landscape writing inspired by and attentive to 'the idea that representations initiate and provoke rather than constrain and tie down; the idea that landscape is not encoded with meaning but is a surplus of meaningfulness' (*ibid.*, p.550).

One clear fact emerges from the work of geographers such as Lorimer, Dubow and Rose. To speak of 'landscape writing' is to reintroduce once more questions of subjectivity and the self – questions concerning the figure who writes, narrates and perceives. In this way, the continuing trace of the subject, of subjects, howsoever ghostly, or embodied, relational and contingent, is a hallmark of recent landscape writing. For Rose and Wylie (2006, p.477), for example, writing about the place of landscape in the wake of non-representational and vitalist theories, notions of subjectivity, perception and narrative represent something of a saving grace:

> Perhaps this is where landscape might creatively insinuate itself into vitalist, relational, and topological geographies: landscape reintroduces perspective and contour; texture and feeling; perception and imagination. It is the synthesis of elements, so elegantly traced by topologies, with something added: lightless chasms, passing clouds, airless summits, sweeping sands. While the introduction of such visions necessarily introduces the spectre of a self-regarding subject (constituted, positioned, and/or constructed in relation to a realised object), it is a spectre that cannot be escaped simply by attending to the microtopological relations constituting these effects. This is why the problem, the tension, of landscape is a tension of presence. It is the tension of regarding at a distance that which enables one to see.

The argument here is that landscape writing potentially supplies a dimension that the topological and hybrid geographies discussed in the previous section lack – a certain depth provided by the perspective of a subject. The counter-argument, of course, from a post-humanist position, is that this runs the risk of lapsing once more into humanist modes of thought that not only privilege human perspectives over others, but also invariably understands human identities as pre-given, stable and coherent. However, as the quote above demonstrates, perhaps the defining feature of recent landscape writing by UK-based cultural geographers is that it is very much written in the light of both phenomenological understanding of the self as embodied and of-the-world and poststructural understandings of selfhood as contingent, fractured, multiple and in various ways historically and culturally constituted. The accent, in all the work discussed in this section, is therefore upon how different senses of self and landscape are emergent and changeable through practices such as writing.

One way of exploring this sense of self and landscape as emergent is through affect – a term and concept that has come to prominence in cultural geography and cultural and political theory more widely since the late 1990s. In cultural geography in particular this has occurred within the ambit of non-representational theories, and the understanding of affect at work here is primarily derived from the work of Deleuze and Guattari (1994). For Deleuze and Guattari, far from being synonymous with emotion (as something felt and experienced by an individual, on the inside), the term affect denotes a non- or pre-subjective intensity. In contemporary human geography, most commonly, affect thus signals the non-rational or more-than-rational aspects of life, and also the broader notion of a charged background of affective capacities and tensions acting as a catalyst for corporeal practice and performance. In this sense affect denotes the shifting mood, tenor, colour or intensity of places and situations. In a deeper sense, however, affect may be understood as 'that through which subject and object emerge and become possible' (Dewsbury et al., 2002, p. 439). This definition highlights that while affect of course implies and involves human emotions, perceptions and sensations, it is not simply reducible to them. Affect precisely describes domains of experience that are more-than subjective and yet at the same time formative of senses of self – and formative, for example, of senses of self and landscape.

This is the argument developed in my own work on coastal walking, perception and narrative (Wylie, 2005), which is intended as a contribution to current explorations of the possibilities of forms of creative landscape writing:

> In the context of coastal walking [affect] connotes configurations of motion and materiality – of light, colour, morphology and mood – *from which* distinctive senses of self and landscape, walker and ground, observer and observed, distil and refract. . . . The circulation and upsurge of affects and percepts is precisely that from which these two horizons, inside and outside, self and landscape, precipitate and fold.
>
> (Wylie, 2005, p.236)

This paper thus seeks to work though various situations encountered on the coast walk – in particular instances of tactile engagement and distanced gazing – in an effort to phenomenologically and experientially describe the emergence of distinct senses of self and landscape. And, though so many have already been given and discussed throughout this book, this involves elaborating another definition of landscape. Landscape is that with which we see:

> Landscape is not just a way of seeing, a projection of cultural meaning. Nor, of course, is landscape simply something seen, a mute, external field. Nor, finally, can we speak altogether plausibly of the practice of self and landscape through notions of a phenomenological *milieu* of dwelling. Taking a first step past constructivist, realist and phenomenological visions, this paper writes its way through what might be termed a post-phenomenological understanding of the formation and undoing of self and landscape in practice. Therein, landscape might best be described in terms of the entwined materialities and sensibilities *with which* we act and sense.
>
> (Wylie, 2005, p.245)

## 6.6 CONCLUSION: CREATIVE TENSIONS

This final chapter has sought to identify and document some of the main conceptual and substantive issues currently being explored by cultural geographers. In turn, given the nascent nature of some of these issues and forms of writing, it is possible to argue that we can take from this some hints as to how landscape studies in geography will develop in the years ahead. Alternatively, of course, some entirely new emphasis may well emerge, a different definition, a different set of topical issues. This chapter thus makes no definite claims regarding the future. But it is in some ways a petition on behalf of landscape, in so far as it has sought to illustrate the continuing vitality and diversity of writing in this area. In this context, it should be noted that, following Mitchell's (2003b) final piece on work on landscape and justice, the journal *Progress in Human Geography* – which publishes regular essays discussing new work in various branches of human geography – opted to discontinue reports on landscape. These had been a feature of the journal since the early 1990s, reflecting, as discussed in Chapters 3 and 4 of this book, the prominence of the landscape concept within geographies of the cultural turn. With the natural ebbing of that particular disciplinary tide, it may have seemed that landscape was also a concept in partial retreat, albeit from an advanced position. But even after only four years have passed, the decision to discontinue landscape reports can be strongly challenged, given the wealth of new writing inspired by phenomenological and non-representational understandings of embodiment, materiality and performance (and this work is of course being highlighted in a new series of cultural geography reports (Lorimer, 2005)). Hopefully the work discussed in Chapter 5 and this chapter amplifies the fact that landscape continues to be, as this series implies, a key idea in geography.

At the start of this book I argued that landscape could perhaps best be thought of as a series of tensions: tensions between distance and proximity, observing and inhabiting, eye and land, culture and nature; these tensions animate the landscape concept, make it cogent and productive. The book as a whole has sought to show how different understandings of landscape inspire different types of research and writing, and this chapter has been no exception. Here we have seen landscape examined in terms of conflicts over memory and identity, conflicts in which landscape emerges as a matter of everyday life *and* an ideological vehicle (Section 6.2). Equally, Kenneth Olwig's (2002) historical and lexicological re-casting of landscape in terms of custom, polity and law charts the long-evolving tensions between landscape as commonplace and as scenic space (Section 6.3). Finally, Sections 6.4 and 6.5 have worked through tensions between topology and topography, and have shown that if, on the one hand, relational and vitalist thinking elaborates a new, hybrid and post-humanist geography for which traditional notions of landscape seem stale, then, on the other, cultural geographers inspired by similar conceptual resources are exploring critical and creative forms of landscape writing, in which ideas of movement, subjectivity, perception and narrative are deployed in nuanced fashion.

It thus seems appropriate to end on a personal note. An emphasis upon personal experience, and upon the tension that obtains between self and landscape, the tension of distance and proximity, was present in a statement referred to on the very first page of the book – the artist Paul Cezanne's declaration that 'the landscape thinks itself in me, and I am its consciousness'. Back then, I noted the egoism lurking in this declaration, in which Cezanne in a way claims the landscape as his, and signs himself as its presiding spirit and animus. But this can be turned around: it is the painter to whom the landscape gives birth; as Maurice Merleau-Ponty says, the painter emerges as a watching, moving, rapt subject in a tension spun through and with the lines of the landscape. Landscape, in other words, is a perceiving-with, that with which we see, the creative tension of self and world.

# BIBLIOGRAPHY

Abrams, D. (1996) *The Spell of the Sensuous*, London: Vintage.

Adams, P., Hoelscher, S. and Till, K. (eds) (2001) *Textures of Place: Exploring Humanist Geographies*, Minneapolis: University of Minnesota Press.

Agamben, G. (1998) *Homo Sacer: Sovereign Power and Bare Life* Stanford, CA: Stanford University Press.

Aitken, S. and Valentine, G. (eds) (2006) *Approaches to Human Geography*, London: Sage.

Alderman, D. (2002) 'Street names as memorial arenas: the reputational politics of commemorating Martin Luther King Jr. in a Georgia county', *Historical Geography* 20: 99–120.

Alpers, S. (1983) *The Art of Describing: Dutch Art in the 17th Century*, London: John Murray.

Anderson, B. (2004) 'Time-stilled, space-slowed: how boredom matters', *Geoforum* 35: 739–754.

Andrews, M. (1984) *The Search for the Picturesque: Landscape Aesthetics and Tourism in Britain, 1760–1800*, Aldershot: Scholar.

Ansell Pearson, K. (1999) *Germinal Life: The Difference and Repetition of Deleuze*, London: Routledge.

Ashcroft, B., Griffiths, G. and Tiffin, H. (eds) (1995) *The Post-Colonial Studies Reader*, London: Routledge.

Ashfield, A. and de Bolla, P. (eds) (1996) *The Sublime: A Reader in British Eighteenth-century Aesthetic Theory*, Cambridge: Cambridge University Press.

Bachelard, G. (1994 [1969]) *The Poetics of Space*, Boston, MA: Beacon Press.

Barnes, T.J. and Duncan, J.S. (1992) 'Introduction: writing worlds', in Barnes, T.J. and Duncan, J.S. (eds) *Writing Worlds*, London: Routledge.

Barnett, C. (1998) 'Impure and worldly geography: the Africanist discourse of the Royal Geographical Society, 1831–1873', *Transactions of the Institute of British Geographers* NS 23(2): 239–251.

—— (1999) 'Deconstructing context: exposing Derrida', *Transactions of the Institute of British Geographers* NS 24: 277–293.

—— (2001) 'Culture, geography, and the arts of government', *Environment and Planning D: Society and Space* 1: 7–24.

Barrell, J. (1980) *The Dark Side of the Landscape*, Cambridge: Cambridge University Press.

Barry, A., Osborne, T. and Rose, N. (eds) (1996) *Foucault and Political Reason: Liberalism, Neo-Liberalism and Rationalities of Government*, London: UCL Press.

Barthes, R. (1977) *Image, Music, Text*, London: Fontana Books.

—— (1990 [1957]) *Mythologies*, Harmondsworth: Penguin.

Battisti, G. (1981) *Brunelleschi*, London: Thames and Hudson.

Baxandall, M. (1988 [1972]) *Painting and Experience in 15th Century Italy: A Primer in the Social History of Pictorial Style*, Oxford: Oxford University Press.

Bender, B. (1994) 'Stonehenge – contested landscapes (medieval to present day)', in Bender, B. (ed.) *Landscape: Politics and Perspectives*, London: Routledge.

Benjamin, W. (1992) *Illuminations*, London: Fontana.

Bennett, J. (2001) *The Enchantment of Modern Life: Attachments, Crossings and Ethics*, Oxford: Berg.

Berger, J. (1972) *Ways of Seeing*, Harmondsworth: Penguin.

Bermingham, A. (1986) *Landscape and Ideology: The English Rustic Tradition, 1740–1860*, Berkeley: University of California Press.

Bhabha, H. K. (1985) 'Signs taken for wonders: questions of ambivalence and authority under a tree outside Delhi, May 1817', *Critical Inquiry* 12(1), 144–165.

Bingham, N. (1996) 'Objections: from technological determinism towards geographies of relations', *Environment and Planning D: Society and Space* 14: 635–657.

—— (2006) 'Bees, butterflies, and bacteria: biotechnology and the politics of nonhuman friendship', *Environment and Planning A* 38(3): 483–498.

Birtles, T.G. (1997) 'First contact: colonial European preconceptions of tropical Queensland rainforest and its people', *Journal of Historical Geography* 23: 393–417.

Blaut, J. (1993) *The Colonizer's Model of the World*, New York: Guilford Press.

Blunt, A. (1994) *Travel, Gender and Imperialism: Mary Kingsley and West Africa*, New York: Guilford Press.

Blunt, A. and McEwan, C. (eds) (2002) *Postcolonial Geographies*, London: Continuum.

Bondi, L. (1992) 'Gender symbols and urban landscapes', *Progress in Human Geography* 16(2): 157–170.

—— (1998) 'Gender, class and urban space: public and private space in contemporary urban landscape', *Urban Geography* 19(2): 160–185.

Bourdieu, P. (1977) *Outline of a Theory of Practice*, Cambridge: Cambridge University Press.

—— (1984) *Distinction: A Social Critique of the Judgement of Taste*, London: Routledge.

Brace, C. (2000) 'A pleasure ground for the noisy herds? Incompatible encounters with the Cotswolds and England, 1900–1950', *Rural History: Economy, Society, Culture* 11(1): 75–94.

Brassley, P. (1998) 'On the unrecognized significance of the ephemeral landscape', *Landscape Research* 23(2): 119–132.

Braun, B. (2000) 'Producing vertical territory: geology and governmentality in late Victorian Canada', *Ecumene* 7: 7–46.

Bravo, M. (1999) 'Precision and curiosity in scientific travel: James Rennell and the orientalist geography of the new imperial age, 1760–1830', in Elsner, J. and Rubies, J-P. (eds) *Voyages and Visions: Towards a Cultural History of Travel*, London: Reaktion.

Bruck, J. (2005) 'Experiencing the past? The development of a phenomenological archaeology in British prehistory', *Archaeological Dialogues* 12(1): 45–72.

Bryant, R. (2001) 'Political ecology: a critical agenda for change?', in Castree, N. and Braun, B. (eds) *Social Nature: Theory, Practice, and Politics*, Oxford: Blackwell.

Buck-Morss, S. (1989) *Dialectics of Seeing: Walter Benjamin and the Arcades Project*, London: MIT Press.

Burchell, G., Gordon, C. and Miller, P. (eds) (1991) *The Foucault Effect: Studies in Governmentality*, Chicago, IL: University of Chicago Press.

Burgin, V. (1996) *In/Different Spaces: Place and Memory in Visual Culture*, Berkeley: University of California Press.

Büscher, M. (2006) 'Vision in motion', *Environment and Planning A* 38(2): 281–299.

Butler, J. (1990) *Gender Trouble: Feminism and the Subversion of Identity*, London: Routledge.

—— (1993) *Bodies that Matter: On the Discursive Limits of Sex*, London: Routledge.

Buttimer, A. (1976) 'Grasping the dynamism of lifeworld', *Annals of the Association of American Geographers* 66: 277–292.

Calaresu, M. (1999) 'Looking for Virgil's tomb: the end of the Grand Tour and the cosmopolitan ideal in Europe', in Elsner, J. and Rubies, J-P. (eds) *Voyages and Visions: Towards a Cultural History of Travel*, London: Reaktion.

Cant, S. G. (2006) 'British speleologies: geographies of science, personality and practice, 1935–1953', *Journal of Historical Geography* 32(4): 775–795.

Carter, P. (1987) *The Road to Botany Bay: An Essay in Spatial History*, London: Faber and Faber.

Casey, E. (1991) '"The element of voluminousness": depth and place re-examined', in Dillon, M.C. (ed.) *Merleau-Ponty Vivant*, Albany, NY: SUNY Press.

—— (1998) *The Fate of Place*, Berkeley: University of California Press.

Castree, N. (2005) *Nature*, London: Routledge.

Castree, N. and Braun, B. (eds) (2001) *Social Nature: Theory, Practice and Politics*, Oxford: Blackwell.

Castree, N. and MacMillan, T. (2001) 'Dissolving dualisms: actor-networks and the reimagination of nature', in Castree, N. and Braun, B. (eds) *Social Nature: Theory, Practice, and Politics*, Oxford: Blackwell.

Chakrabarty, D. (1992) 'Postcoloniality and the artifice of history', in Ashcroft, B., Griffiths, G. and Tiffin, H. (eds) *The Post-colonial Studies Reader*, London: Routledge.

Chard, C. (1999) *Pleasure and Guilt on the Grand Tour: Travel Writing and Imaginative Geography, 1600–1830*, Manchester: Manchester University Press.

Cherry-Garrad, A. (1939) *The Worst Journey in the World*, London: Chatto and Windus.

Clark, K. (1969) *Civilisation: A Personal View*, London: John Murray.

Clark, N. (2002) 'The demon-seed: bioinvasion as the unsettling of environmental cosmopolitanism', *Theory Culture and Society* 19(1–2): 101–126.

Clayton, D. (2000a) *Islands of Truth*, Vancouver: University of British Columbia Press.

—— (2000b) 'On the colonial genealogy of George Vancouver's chart of the north-west coast of North America', *Ecumene* 7: 371–401.

—— (2000c) 'The creation of imperial space in the Pacific Northwest', *Journal of Historical Geography* 26: 327–350.

Clifford, J. and Marcus, G. (eds) (1986) *Writing Culture: The Poetics and Politics of Ethnography*, Berkeley: University of California Press.

Cloke, P. and Jones, O. (2001) 'Dwelling, place and landscape: an orchard in Somerset', *Environment and Planning A* 33: 649–666.

Cloke, P., Philo, C. and Sadler, D. (1991) *Approaching Human Geography*, London: Sage.

Colebrook, C. (2002) *Deleuze*, London: Routledge.

Connolly, W. (2002) *Neuropolitics: Thinking, Culture, Speed*, Minneapolis: University of Minnesota Press.

Conradson, D. (2005) 'Landscape, care and the relational self: therapeutic encounters in rural England', *Health and Place* 11: 337–348.

Cosgrove, D. (1983) 'Towards a radical cultural geography', *Antipode* 15: 1–11.

—— (1985) 'Prospect, perspective and the evolution of the landscape idea', *Transactions of the Institute of British Geographers* NS 10(1): 45–62.

—— (1994) 'Contested global visions: one-world, whole-earth and the Apollo space photographs', *Annals of the Association of American Geographers* 84(2): 270–294.

—— (1998 [1984]) *Social Formation and Symbolic Landscape*, Madison: University of Wisconsin Press.

—— (ed.) (1999) *Mappings*, London: Reaktion.

Cosgrove, D. and Thornes, J. (1981) 'Of truth of clouds', in Pocock, D.C. (ed.) *Humanistic Geography and Literature*, London: Croom Helm.

Cosgrove, D. and Jackson, P. (1987) 'New directions in cultural geography', *Area* 19: 95–101.

Cosgrove, D. and Daniels, S. (1988a) 'Introduction', in Daniels, S. and Cosgrove, D. (eds) *The Iconography of Landscape*, Cambridge: Cambridge University Press.

—— (eds) (1988b) *The Iconography of Landscape*, Cambridge: Cambridge University Press.

Cosgrove, D. and Duncan, J. (1995) 'Editorial', *Ecumene* 2(2): 127–128.

Crary, J. (1990) *Techniques of the Observer: On Vision and Modernity in the 19th Century*, Cambridge, MA: MIT Press.

—— (2000) *Suspensions of Perception: Attention, Spectacle and Modern Culture*, Cambridge MA: MIT Press

Cresswell, T. (1996) *In Place/Out of Place: Geography, Ideology and Transgression*, Minneapolis: University of Minnesota Press.

—— (2000) 'Falling down: resistance as diagnostic', in Sharp, J.P., Routledge, P., Philo, C. and Paddison, R. (eds) *Entanglements of Power: Geographies of Domination/Resistance* London: Routledge.

—— (2002) 'Bourdieu's geographies: in memorium', *Environment and Planning D: Society and Space* 20: 379–383.

—— (2003) 'Landscape and the obliteration of practice', in Anderson, K., Domosh, D., Pile, S. and Thrift, N. (eds) *Handbook of Cultural Geography*, London: Sage.

Crossley, N. (1995) 'Merleau-Ponty, the elusive body and carnal sociology', *Body and Society* 1(1): 43–63.

—— (1996) 'Body-subject/body-power: agency, inscription and control in Foucault and Merleau-Ponty', *Body and Society* 2(2): 99–116.

Crouch, D. (2003) 'Spacing, performing and becoming: tangles in the mundane', *Environment and Planning A* 35: 1,945–1,960.

Damisch, H. (1994) *The Origin of Perspective*, Cambridge, MA: MIT Press.

Daniels, S. (1982) 'Humphrey Repton and the morality of landscape', in Gold, J. and Burgess, J. (eds) *Valued Environments*, London: Allen and Unwin.

—— (1985) 'Arguments for a humanistic geography', in Johnston, R.J. (ed.) *The Future of Geography*, London: Methuen.

—— (1989) 'Marxism, culture and the duplicity of landscape', in Peet, R. and Thrift, N. (eds) *New Models in Geography*, Vol. 2, London: Unwin Hyman.

—— (1993) *Fields of Vision: Landscape Imagery and National Identity*, Cambridge: Polity Press.

Darby, W.J. (2000) *Landscape and Identity: Geographies of Nation and Class in England*, Oxford: Berg.

De Bolla, P. (1989) *The Discourse of the Sublime*, Oxford: Basil Blackwell.

De Certeau, M. (1983) 'The madness of vision', *Diacritics* 13: 24–31.

—— (1984) *The Practice of Everyday Life*, Berkeley: University of California Press.

DeLyser, D. (1999) 'Authenticity on the ground: engaging the past in a California ghost town', *Annals of the Association of American Geographers* 89(4): 602–632.

DeSilvey, C. (2003) 'Cultivated histories in a Scottish allotment garden', *Cultural Geographies* 10: 442–468.

Degen, M., Whatmore, S., Hinchliffe, S. and Kearnes, M.B. (2006, forthcoming) 'Greenfingers, greenspaces: passionate involvements with urban natures', *Body and Society*.

Deleuze, G. (1988a) *Cinema 1: The Movement-Image*, London: Athlone Press.

—— (1988b) *Foucault*, London: Athlone Press.

—— (1990) *The Logic of Sense*, London: Athlone Press.

—— (1992) *Expressionism in Philosophy: Spinoza*, London: Zone Books.

—— (1994a) *The Fold: Leibniz and the Baroque*, Minneapolis: University of Minnesota Press.

—— (1994b) *Difference and Repitition*, London: Athlone Press.

—— (2001) *Pure Immanence*, New York: Zone Books.

Deleuze, G. and Guttari, F. (1988) *A Thousand Plateaus: Capitalism and Schizophrenia*, London: Athlone Press.

—— (1994) *What is Philosophy?* London: Verso.

Demeritt, D. (1994) 'The nature of metaphors in cultural geography and environmental history', *Progress in Human Geography* 18(2): 163–185.

Derrida, J. (1973) *Speech and Phenomena, and Other Essays on Husserl's Theory of Signs*, Evanston, IL: Northwestern University Press.

—— (1976) *Of Grammatology*, London: Johns Hopkins University Press.

—— (1978) *Writing and Difference*, London: Routledge.

—— (1982) *Margins of Philosophy*, Brighton: Harvester Press.

Descartes, R. (1975 [1641]) *Philosophical Writings*, edited and selected by E. Anscombe and P.T. Geach, London: Nelson.

Dewsbury, J-D. (2000) 'Performativity and the event: enacting a philosophy of difference', *Environment and Planning D: Society and Space* 18: 473–496.

—— (2003) 'Witnessing space: knowledge without contemplation', *Environment and Planning A* 35(11): 1,907–1,933.

Dewsbury, J-D., Wylie, J., Harrison, P. and Rose, M. (2002) 'Enacting geographies', *Geoforum* 32: 437–441.

Dillon, M.C. (1997) *Merleau-Ponty's Ontology*, Evanston, IL: Northwestern University Press.

Dodds, K. (1996) 'To photograph the Antarctic: British Polar exploration and the Falklands Islands and Dependencies Aerial Survey Expeditions', *Ecumene* 3: 63–90.

Domosh, M. (1996) *Invented Cities: The Creation of Landscape in Nineteenth-Century New York and Boston*, New Haven, CT: Yale University Press.

Dreyfus, H. (1991) *Being-in-the-World: A Commentary on Heidegger's Being and Time, Division 1*, London: MIT Press.

Dreyfus, H. and Rabinow, P. (1982) *Michel Foucault: Beyond Structuralism and Hermeneutics*, Berkeley: University of California Press.

Driver, F. (1992) 'Geography's empire: histories of geographical knowledge', *Environment and Planning D: Society and Space* 10: 23–40.

—— (1995) 'Geographical traditions: rethinking the history of geography', *Transactions of the Institute of British Geographers* NS 20(4): 403–405.

—— (2001) *Geography Militant*, Oxford: Blackwell.

Driver, F. and Gilbert, D. (1998) 'Heart of Empire: landscape, space and performance in imperial London', *Environment and Planning D: Society and Space* 16: 11–28.

Driver, F. and Martins, L. (eds) (2005) *Tropical Visions in an Age of Empire*, Chicago, IL: University of Chicago Press.

Dubow, J. (2000) 'From a view on the world to a point of view in it: rethinking sight, space and the colonial subject', *Interventions: International Journal of Postcolonial Studies* 2(1): 87–102.

—— (2001) 'Rites of passage: travel and the materiality of vision at the Cape of Good Hope', in Bender, B. and Winer, M. (eds) *Contested Landscapes: Movement, Exile and Place*, Oxford: Berg.

—— (2004) 'The mobility of thought: reflections on Blanchot and Benjamin', *Interventions: International Journal of Postcolonial Studies* 6(2): 216–228.

Duncan, J. (1980) 'The superorganic in American cultural geography', *Annals of the Association of American Geographers* 70(2): 181–198.

—— (1990) *The City as Text: The Politics of Landscape Interpretation in the Kandyan Kingdom*, Cambridge: Cambridge University Press.

—— (1993) 'Sites of representation: place, time and the discourse of the other', in Duncan, J. and Ley, D. (eds) *Place/Culture/Representation*, London: Routledge.

—— (1995) 'Landscape geography, 1993–1994', *Progress in Human Geography* 19(3): 414–422.

—— (1999) 'Dis-orientation: on the shock of the familiar in a far away place', in Duncan, J. and Gregory, D. (eds) *Writes of Passage: Reading Travel Writing*, London: Routledge.

Duncan, J. and Duncan, N. (1988) '(Re)reading the landscape', *Environment and Planning D: Society and Space* 6: 117–126.

—— (1992) 'Ideology and bliss: Roland Barthes and the secret histories of landscape', in Barnes, T.J. and Duncan, J.S. (eds) *Writing Worlds*, London: Routledge.

Duncan, J. and Ley, D. (1993) 'Introduction: representing the place of culture', in Duncan, J. and Ley, D. (eds) *Place/Culture/Representation*, London: Routledge.

Duncan, J. and Gregory, D. (eds) (1999) *Writes of Passage: Reading Travel Writing*, London: Routledge.

Duncan, J. and Duncan, N. (2003) 'Can't live with them; can't landscape without them: racism and the pastoral aesthetic in suburban New York', *Landscape Journal* 22(1): 88–98.

—— (2004) *Landscapes of Privilege: The Politics of the Aesthetic in an American Suburb*, London: Routledge.

Dwyer, O.J. (2000) 'Interpreting the Civil Rights Movement: place, memory, and conflict', *Professional Geographer* 52: 660–671.

—— (2004) 'Symbolic accretion and commemoration', *Social and Cultural Geography* 5(3): 419–435.

Eagleton, T. (1983) *Literary Theory: An Introduction*, Oxford: Basil Blackwell.

Eden, S., Tunstall, S. and Tapsell, S. (2000) 'Translating nature: river restoration and nature–culture', *Environment and Planning D: Society and Space* 14: 257–273.

Edensor, T. (2000) 'Walking in the countryside: reflexivity, embodied practices and ways to escape', *Body and Society* 6: 81–106.

Edgerton, S.Y. (1975) *The Renaissance Rediscovery of Linear Perspective*, London: Harper and Row.

Eliade, M. (1973) *The Sacred and the Profane*, New York: Harper Row.

Elkins, J. (1992) 'Multiple perspectives' *Journal of the History of Ideas* 53(2): 209–230.

—— (1994) *The Poetics of Perspective*, Ithaca, NY: Cornell University Press.

Elsner, J. and Rubies, J-P. (eds) (1999) *Voyages and Visions: Towards a Cultural History of Travel*, London: Reaktion.

Fanelli, G. (1977) *Brunelleschi*, Florence: Scala Institutio Fotographica Editoriale.

Ford, S. (1991) 'Landscape revisited: a feminist reappraisal', in Philo, C. (compiler) *New Words, New Worlds: Reconceptualising Social and Cultural Geography*, Lampeter, Social and Cultural Geography Study Group of the Institute of British Geographers.

Foster, J. (1998) 'John Buchan's "Hesperides": landscape rhetoric and the aesthetics of bodily experience on the South African Highveld, 1901–1903', *Ecumene* 5(3): 323–347.

Foucault, M. (1977) *The Order of Things*, London: Routledge.

—— (1981 [1976]) *The History of Sexuality, Vol. 1: An Introduction*, London: Allen Lane.

—— (1982a) 'Technologies of the self', in Martin, L.H., Gutman, H. and Hutton, P.H. (eds) *Technologies of the Self: A Seminar with Michel Foucault*, Amherst: University of Massachusetts Press

—— (1982b) 'The subject and power', *Critical Inquiry* 8(4): 777–795.

—— (1989a [1961]) *Madness and Civilization: A History of Insanity in the Age of Reason*, London: Routledge.

—— (1989b [1969]) *The Archaeology of Knowledge*, London: Routledge.

—— (1991a [1975]) *Discipline and Punish: The Birth of the Prison*, London: Penguin.

—— (1991b) 'Governmentality', in Burchell, G., Gordon, C. and Miller, P. (eds) *The Foucault Effect: Studies in Governmentality*, Chicago, IL: University of Chicago Press.

—— (1991c) *The Foucault Reader*, edited by Paul Rabinow, London: Penguin.

—— (1992a [1984]) *The History of Sexuality, Vol. 2: The Use of Pleasure*, London: Penguin.

—— (1992b [1986]) *The History of Sexuality, Vol. 3: The Care of the Self*, London: Penguin.

—— (1997) *Ethics: Subjectivity and Truth*, London: Allen Lane.

Geertz, C. (1973) *The Interpretation of Cultures*, New York: Basic Books.

Gesler, W. and Kearns, R. (eds) (1998) *Putting Health into Place: Landscape, Identity and Wellbeing*, Syracuse, NY: Syracuse University Press.

Gibson, J.J. (1950) *The Perception of the Visual World*, Boston, MA: Houghton Mifflin.

—— (1979) *The Ecological Approach to Visual Perception*, Boston, MA: Houghton Mifflin.

Gillies, J. (1994) *Shakespeare and the Geography of Difference*, Cambridge: Cambridge University Press.

Gilroy, A. (ed.) (2000) *Romantic Geographies: Discourses of Travel, 1775–1844*, Manchester: Manchester University Press.

Gold, J. and Burgess, J. (eds) (1982) *Valued Environments*, London: Allen and Unwin.

Green, N. (1990) *The Spectacle of Nature: Landscape and Bourgeois Culture in 19th-Century France*, Manchester: Manchester University Press.

Greenblatt, S. (1991) *Marvellous Possessions: The Wonder of the New World*, Oxford: Clarendon Press.

Greenhough, B. and Roe, E. (2006) 'Towards a geography of bodily bio-technologies', *Environment and Planning A* 38: 416–422.

Gregory, D. (1994) *Geographical Imaginations*, Oxford: Basil Blackwell.

—— (1995) 'Between the book and the lamp: imaginative geographies of Egypt, 1849–50', *Transactions of the Institute of British Geographers* 20: 29–57.

—— (1999) 'Scripting Egypt: Orientalism and the cultures of travel', in Duncan, J. and Gregory, D. (eds) *Writes of Passage: Reading Travel Writing*, London: Routledge.

Gregory, D. and Duncan, J. (eds) (1999) *Writes of Passage: Reading Travel Writing*, London: Routledge.

Gregson, N. and Rose, G. (2000) 'Taking Butler elsewhere: performativities, spatialities and subjectivities', *Environment and Planning D: Society and Space* 18: 433–452.

Grosz, E. (1994) *Volatile Bodies: Towards a Corporeal Feminism*, Bloomington: Indiana University Press.

Gruffudd, P. (1996) 'The countryside as educator: schools, rurality and citizenship in inter-war Wales', *Journal of Historical Geography* 22: 412–423.

Hamilakis, Y., Pluciennik, M. and Tarlow, S. (2002) 'Introduction: thinking through the body', in Hamilakis, Y., Pluciennik, M. and Tarlow, S. (eds) *Thinking through the Body: Archaeologies of Corporeality*, London: Routledge.

Haraway, D. (1991) *Simians, Cyborgs and Women: The Reinvention of Nature*, London: Routledge.

Harrer, S. (2005) 'The theme of subjectivity in Foucault's lecture series *L'Herméneutique du Sujet*', *Foucault Studies* 2: 75–96.

Harrison, P. (2000) 'Making sense: embodiment and the sensibilities of the everyday', *Environment and Planning D: Society and Space* 18: 497–517.

—— (2007) '"How shall I say it . . .?" Relating the non-relational', *Environment and Planning A* 39: 590–608.

—— (2007, forthcoming) 'The space between us. Opening remarks on the concept of dwelling', *Environment and Planning D: Society and Space*.

Harrison, S., Pile, S. and Thrift, N. (eds) (2004) *Patterned Ground: Entanglements of Nature and Culture*, London: Reaktion.

Hartshone, R. (1939) 'The nature of geography: a critical survey of current thought in the light of the past', *Annals of the Association of American Geographers* 29(3–4).

Harvey, D. (1996) *Justice, Nature and the Geography of Difference*, Oxford: Blackwell.

Heffernan, M. (1995) 'For ever England: the Western Front and the politics of remembrance in Britain', *Ecumene* 2: 293–324.

—— (2001) '"A dream as frail as those of ancient time": the incredible geographies of Timbuctoo', *Environment and Planning D: Society and Space* 19: 203–225.

Heidegger, M. (1962) *Being and Time*, Oxford: Basil Blackwell.

—— (1996) 'Building dwelling thinking', in *Basic Writings* (ed. D. Krell), London: Routledge.

Henderson, G. (2003) 'What (else) we talk about when we talk about landscape: for a return to the social imagination', in Wilson, C. and Groth, P. (eds) *Everyday America: Cultural Landscape Studies after J.B. Jackson*, Berkeley: University of California Press.

Hetherington, K. (1997) 'In place of geometry: the materiality of place', in Hetherington, K. and Munro, R. (eds) *Ideas of Difference: Social Spaces and the Labour of Division*, Oxford: Basil Blackwell.

—— (2003) 'Spatial textures: place, touch and *praesentia*', *Environment and Planning A* 35: 1933–1944.

Hevly, B. (1996) 'The heroic science of glacier motion', *Osiris* 11: 66–86.

Hewison, R. (1987) *The Heritage Industry*, London: Methuen.

Hinchliffe, S. (2000) 'Entangled humans: specifying powers and their spatialities', in Sharp, J.P., Routledge, P., Philo, C. and Paddison, R. (eds) *Entanglements of Power: Geographies of Domination/Resistance* London: Routledge.

—— (2001) 'Indeterminacy in-decisions: science, policy and politics in the BSE crisis', *Transactions of the Institute of British Geographers* 26(2): 182–204.

—— (2003) 'Inhabiting – landscapes and natures', in Anderson, K., Domosh, M., Pile, S. and Thrift, N. (eds) *Handbook of Cultural Geography*, London: Sage.

Hinchliffe, S., Kearnes, M.B., Degen, M. and Whatmore, S. (2005) 'Urban wild things – a cosmopolitical experiment', *Environment and Planning D: Society and Space* 23(5): 643–658.

Hitchings, R. (2003) 'People, plants and performance: on actor network theory and the material pleasures of the private garden', *Social and Cultural Geography* 4: 99–113.

Hoelscher, S. and Alderman, D. (2004) 'Memory and place: geographies of a critical relationship', *Social and Cultural Geography* 5(3): 347–355.

Holloway, J. (2003) 'Make-believe: spiritual practice, embodiment, and sacred space', *Environment and Planning A* 35(11): 1,961–1,974.

Hoskins, W.G. (1970) *The Shell Guide to Leicestershire*, London: Faber.

—— (1985 [1954]) *The Making of the English Landscape*, London: Penguin.

Hughes, A. and Reimer, S. (eds) (2004) *Geographies of Commodity Chains*, London: Routledge.

Ingold, T. (1993) 'The temporality of the landscape', *World Archaeology* 25(2): 152–171. Reprinted in Ingold, T., *The Perception of the Environment*, London: Routledge, 2000.

—— (1995) 'Building, dwelling, living: how people and animals make themselves at home in the world', in Strathern, M. (ed.) *Shifting Contexts: Transformations in Anthropological Knowledge*, London: Routledge.

—— (2000) *The Perception of the Environment: Essays on Livelihood, Dwelling and Skill*, London: Routledge.

—— (2005) 'The eye of the storm: visual perception and the weather', *Visual Studies* 20(2): 97–104.

Jackson, J.B. (1984a) 'Agrophilia, or, the love of horizontal spaces', in *Discovering the Vernacular Landscape*, New Haven, CT: Yale University Press.

—— (1984b) *Discovering the Vernacular Landscape*, New Haven, CT: Yale University Press.

—— (1995) 'In search of the proto-landscape', in Thompson, G.F. (ed.) *Landscape in America*, Austin: University of Texas Press.

—— (1997a [1957–58]) 'The Abstract World of the Hot-Rodder', in Jackson, J.B. *Landscape in Sight: Looking at America*, New Haven, CT: Yale University Press.

—— (1997b [1960]) *Landscape in Sight: Looking at America*, New Haven, CT: Yale University Press.

Jackson, P. (1989) *Maps of Meaning: An Introduction to Cultural Geography*, London: Unwin Hyman.

Jacobs, M. (1995) *The Painted Voyage: Art, Travel and Exploration 1574–1875*, London: British Museum Press.

Jay, M. (1992) 'Scopic regimes of modernity', in Lash, S. and Friedmann, J. (eds) *Modernity and Identity*, Oxford: Basil Blackwell.

—— (1993) *Downcast Eyes: The Denigration of Vision in 20th Century French Thought*, Berkeley: University of California Press.

Jazeel, T. (2005) 'Nature, nationhood and the poetics of meaning in Ruhuna (Yala) National Park, Sri Lanka', *Cultural Geographies* 12(2): 199–228.

Johnson, M. (2006) *Ideas of Landscape: An Introduction*, Oxford: Blackwell.

Johnson, N.C. (1995) 'Cast in stone: monuments, geography, and nationalism', *Environment and Planning D: Society and Space* 13: 51–65.

Judovitz, D. (1993) 'Vision, representation and technology in Descartes', in Levin, D. (ed.) *Modernity and the Hegemony of Vision*, Berkeley: University of California Press.

Kastner, J. and Wallis, B. (eds) (1998) *Land and Environmental Art*, New York: Phaidon.

Kearns G. (1997) 'The imperial subject: geography and travel in the work of Mary Kingsley and Halford Mackinder', *Transactions of the Institute of British Geographers* 22: 450–472.

Keith, M. and Pile, S. (eds) (1998) *Geographies of Resistance*, London: Routledge.

Kemp, M. (1990) *The Science of Art*, New Haven, CT: Yale University Press.

Lacan, J. (1977) *The Four Fundamental Concepts of Psychoanalysis*, London: Hogarth Press.

Latham, A. and Conradson, D. (2003) 'The possibilities of performance', *Environment and Planning A* 35: 1901–1906.

Latour, B. (1987) *Science in Action: How to Follow Scientists and Engineers Through Society*, Milton Keynes: Open University Press.

—— (1991) 'Technology is society made durable', in Law, J. (ed.) *A Sociology of Monsters: Essays on Power, Technology and Domination*, London: Routledge.

—— (1993) *We Have Never Been Modern*, London: Harvester Wheatsheaf.

—— (1996) *Aramis, or, the Love of Technology*, Cambridge, MA: Harvard University Press.

—— (1998) 'To modernise or ecologise? That is the question', in Braun, B. and Castree, N. (eds) *Remaking Reality: Nature and the Millennium*, London: Routledge.

—— (1999) 'On recalling ANT', in Law, J. and Hassard, J. (eds) *Actor Network Theory and After*, Oxford: Blackwell/Sociological Review.

Laurier, E. and Philo, C. (2004) 'Ethnoarchaeology and undefined investigations', *Environment and Planning A* 36: 421–436.

Law, J. (1994) *Organising Modernity*, Oxford: Basil Blackwell.

—— (1999) 'After ANT: topology, naming and complexity', in Law, J. and Hassard, J. (eds) *Actor Network Theory and After*, Oxford: Blackwell/Sociological Review.

—— (2001) *Aircraft Stories: Decentering the Object in Technoscience*, Durham, NC: Duke University Press.

Law, J. and Mol, A. (1995) 'Notes on materiality and sociality', *Sociological Review* 43: 274–294.

Law, J. and Benchop, R. (1997) 'Resisting narrative euclideanism: Representation, distribution and ontological politics', in Hetherington, K. and Munro, R. (eds) *Ideas of Difference: Social Spaces and the Labour of Division*, Oxford: Basil Blackwell.

Law, J. and Hassard, J. (eds) (1999) *Actor Network Theory and After*, Oxford: Blackwell/Sociological Review.

Leib, J.I. (2002) 'Separate times, shared spaces: Arthur Ashe, Monument Avenue, and the politics of Richmond, Virginia's symbolic landscape', *Cultural Geographies* 9: 286–312.

Lemke, T. (2001) 'The birth of bio-politics – Michel Foucault's lecture at the Collège de France on neo-liberal governmentality', *Economy and Society* 30(2): 190–207.

Levin, D.M. (1990) *The Opening of Vision: Nihilism and the Postmodern Situation*, Evanston, IL: Northwestern University Press.

—— (ed.) (1993) *Modernity and the Hegemony of Vision*, Berkeley: University of California Press.

Levinas, E. (1999 [1969]) *Totality and Infinity*, Pittsburgh, PA: Duquesne University Press.

Lewis, N. (2000) 'The climbing body: nature and the experience of modernity', *Body and Society* 6: 58–81.

Lewis, P (1979) 'Axioms for reading the landscape', in Meinig, D. (ed.) *The Interpretation of Ordinary Landscapes*, Oxford: Oxford University Press.

Linehan, D. (2003) 'Regional survey and the economic geographies of Britain 1930–1939', *Transactions of the Institute of British Geographers* 28(1): 96–122.

Lingis, A. (1968), 'Translator's preface', in Merleau-Ponty, M. *The Visible and the Invisible*, Evanston, IL: Northwestern University Press.

—— (1994) *Foreign Bodies*, Bloomington: Indiana University Press.

—— (1998) *The Imperative*, Bloomington: Indiana University Press.

—— (1999) *Dangerous Emotions*, Bloomington: Indiana Univeristy Press.

Livingstone, D.N. (1992) *The Geographical Tradition: Episodes in the History of a Contested Enterprise*, Oxford: Basil Blackwell.

Lorimer, H. (2000) 'Guns, game and the grandee: the cultural politics of deer-stalking in the Scottish Highlands', *Ecumene* 7(4): 431–459.

—— (2003a) 'Telling small stories: spaces of knowledge and the practice of geography', *Transactions of the Institute of British Geographers* 28: 197–218.

—— (2003b) 'The geographical field course as active archive', *Cultural Geographies* 10: 278–308.

—— (2005) 'Cultural geography: the busyness of being "more-than-representational"', *Progress in Human Geography* 29(1): 83–94.

—— (2006) 'Herding memories of humans and animals', *Environment and Planning D: Society and Space* 24: 497–518.

Lorimer, H. and Lund, K. (2003) 'Performing facts: finding a way through Scotland's mountains', in Szerszynski, B., Heim, W. and Waterton, C. (eds) *Nature Performed: Environment, Culture and Performance*, London: Blackwells.

McCormack, D. (2002) 'A paper with an interest in rhythm', *Geoforum* 33: 469–485.

—— (2003) 'An event of geographical ethics in spaces of affect', *Transactions of the Institute of British Geographers* 28: 458–508.

MacDonald, F. (2001) 'St Kilda and the sublime', *Ecumene* 8: 151–174.

MacKenzie, J. (ed.) (1990) *Imperialism and the Natural World*, Manchester: Manchester University Press.

McNaughten, P. and Urry, J. (1998) *Contested Natures*, London: Sage.

—— (2000) 'Bodies of nature: introduction', *Body and Society* 6: 1–12.

Marks, J. (1998) *Gilles Deleuze: Vitalism and Multiplicity*, London: Pluto Press.

Marston, S., Jones, J.P. and Woodward, K. (2005) 'Human geography without scale', *Transactions of the Institute of British Geographers* 30: 416–432.

Martin, D. and Scherr, A. (2005) 'Lawyering landscapes: lawyers as constituents of landscape', *Landscape Research* 30(3): 379–393.

Martin, L.H., Gutman, H. and Hutton, P.H. (eds) (1982) *Technologies of the Self: A Seminar with Michel Foucault*, Amherst: University of Massachusetts Press.

Martins, L. (1999) 'Mapping tropical waters', in Cosgrove, D. (ed.) *Mappings*, London: Reaktion.

Massey, D. (2004) 'Geographies of responsibility', *Geografiska Annaler B* 86: 5–18.

Massumi, B. (2002) *Parables for the Virtual*, Durham, NC: Duke University Press.

Matless, D. (1992) 'An occasion for geography: landscape, representation and Foucault's corpus', *Environment and Planning D: Society and Space* 10: 41–56.

—— (1993) 'One man's England: W.G. Hoskins and the English culture of landscape', *Rural History* 4(2): 187–207.

—— (1995a) 'Culture run riot? Work in social and cultural geography, 1994', *Progress in Human Geography* 19(3): 395–404.

—— (1995b) 'The art of right living', in Pile, S. and Thrift, N. (eds) *Mapping the Subject: Geographies of Cultural Transformation*, London: Routledge.

—— (1996) 'New material? Work in social and cultural geography, 1995', *Progress in Human Geography* 20(3): 379–392.

—— (1998) *Landscape and Englishness*, London: Reaktion.

—— (2000) 'Action and noise over a hundred years: the making of a nature region', *Body and Society* 6: 141–165.

—— (2003) 'The properties of landscape', in Anderson, K., Domosh, M., Pile, S. and Thrift, N. (eds) *Handbook of Cultural Geography*, London: Sage.

Meinig, D. (1979a), 'Reading the landscape: an appreciation of W.G. Hoskins and J.B. Jackson' in Meinig, D. (ed.) *The Interpretation of Ordinary Landscapes*, Oxford: Oxford University Press.

—— (ed.) (1979b) *The Interpretation of Ordinary Landscapes*, Oxford: Oxford University Press.

Mels, T. (2003) 'Landscape unmasked: Kenneth Olwig and the ghostly relations between concepts', *Cultural Geographies* 10: 379–387.

—— (2005) 'Between "platial" imaginations and Spatial rationalities: navigating justice and law in the low countries', *Landscape Research* 30(3): 321–335.

Merleau-Ponty, M. (1962 [1942]) *Phenomenology of Perception*, London: Routledge and Kegan Paul.

—— (1968 [1961]) *The Visible and the Invisible*, Evanston, IL: Northwestern University Press.

—— (1969a) 'Cezanne's doubt', in Merleau-Ponty, M. *The Essential Writings of Merleau-Ponty*, New York: Harcourt, Brace and World.

—— (1969b) 'Eye and mind', in Merleau-Ponty, M. *The Essential Writings of Merleau-Ponty*, New York: Harcourt, Brace and World.

—— (1969c) 'What is phenomenology?', in Merleau-Ponty, M. *The Essential Writings of Merleau-Ponty*, New York: Harcourt, Brace and World.

Merriman, P. (2005a) '"Respect the life of the countryside": the Country Code, government, and the conduct of visitors to the countryside in post-war England and Wales', *Transactions of the Institute of British Geographers* 30(3): 336–350.

—— (2005b) 'Materiality, subjectification and government: the geographies of Britain's Motorway Code', *Environment and Planning D: Society and Space* 23(2): 235–250.

Michaels, M. (2000) '"These boots were made for walking": mundane technology, the body and human–environment relations', *Body and Society* 6: 107–126.

Miller, D. (ed.) (1997) *Material Cultures*, London: UCL Press.

—— (ed.) (2005) *Materiality*, Durham, NC, and London: Duke University Press.

Mills, S.(1996) 'Gender and colonial space', *Gender, Place and Culture – A Journal of Feminist Geography* 3: 125–148.

Mitchell, D. (1994) 'Landscape and surplus value: the making of the ordinary in Brentwood, CA', *Environment and Planning D: Society and Space* 12: 7–30.

—— (1995) 'There's no such thing as culture: towards a reconceptualisation of the idea of culture in geography', *Transactions of the Institute of British Geographers* 19: 102–116.

—— (1996) *The Lie of the Land: Migrant Workers and the California Landscape*, Minneapolis: University of Minnesota Press.

—— (1998a) 'Writing the Western: new Western history's encounter with landscape', *Ecumene* 5: 7–29.

—— (1998b) *Cultural Geography: A Critical Introduction*, Oxford: Basil Blackwell.

—— (2001) 'The devil's arm: points of passage, networks of violence and the political economy of landscape', *New Formations* 43: 44–60.

—— (2003a) 'California living, California dying: dead labor and the political economy of landscape', in Anderson, K., Pile, S. and Thrift, N. (eds) *Handbook of Cultural Geography*, London: Sage.

—— (2003b) 'Cultural landscapes: just landscapes or landscapes of justice?' *Progress in Human Geography* 27(6): 787–796.

Mitchell, T. (1988) *Colonising Egypt*, Cambridge: Cambridge University Press.

Mitchell, W.J.T. (1994a) 'Imperial landscape', in Mitchell, W.J.T. (ed.) *Landscape and Power*, London: Routledge.

—— (ed.) (1994b) *Landscape and Power* (2nd edition 2002), London: Routledge.

Monk, J. (1992) 'Gender in the landscape: expressions of power and meaning', in Anderson, K. and Gale, F. (eds) *Inventing Places: Studies in Cultural Geography*, London: Longman.

Muir, R. (1998) *Approaches to Landscape*, Basingstoke: Macmillan Press.

—— (2000 [1981]) *The New Reading the Landscape: Fieldwork in Landscape History*, Exeter: University of Exeter Press

Mulhall, S. (1996) *Heidegger and Being and Time*, London: Routledge.

Mulvey, L. (1989) *Visual and Other Pleasures*, Basingstoke: Macmillan.

Murdoch, J. (1997) 'Inhuman/nonhuman/human: actor-network theory and the propects for a nondualistic and symmetrical perspective on nature and society', *Environment and Planning D: Society and Space* 15: 731–756.

—— (1998) 'The spaces of actor network theory', *Geoforum* 29(4): 357–374.

Nash, C. (1996) 'Reclaiming vision: looking at landscape and the body', *Gender Place and Culture* 3: 149–169.

—— (2000) 'Performativity in practice: some recent work in cultural geography', *Progress in Human Geography* 24(4): 653–664.

Naylor, S. (2001) 'Discovering nature, rediscovering the self: natural historians and the landscapes of Agentina', *Environment and Planning D: Society and Space* 19: 227–247.

Naylor, S. and Jones, G. (1997) 'Writing orderly geographies of distant places: the Regional Survey Movement and Latin America', *Ecumene* 3: 454–472.

Neumann, R.P. (1996) 'Dukes, earls and ersatz Edens: aristocratic nature preservationists in colonial Africa', *Environment and Planning D: Society and Space* 14: 79–98.

Norton, W. (2000) *Cultural Geography: Themes, Concepts, Analyses*, Oxford: Oxford University Press.

Olson, E. and Silvey, R. (2006) 'Transnational geographies: rescaling development, migration, and religion', *Environment and Planning A* 38: 805–808.

Olwig, K. (1996) 'Recovering the substantive nature of landscape', *Annals of the Association of American Geographers* 86: 630–653.

—— (2002) *Landscape, Nature and the Body Politic*, Madison: University of Wisconsin Press.

—— (2005a) 'Editorial: law, polity and the changing meaning of landscape', *Landscape Research* 30(3): 293–298.

—— (2005b) 'The landscape of "customary" law versus that of "natural" law', *Landscape Research* 30(3): 299–320.

—— (2005c) 'Representation and alienation in the political land-scape', *Cultural Geographies* 12(1): 19–40.

Outram, D. (1995) *The Enlightenment*, Cambridge: Cambridge University Press.

Panofsky, E. (1991) *Perspective as Symbolic Form*, London: Zone Books.

Pels, D., Hetherington, K. and Vandenberghe, F. (2002) 'The status of the object: performances, mediations and techniques', *Theory, Culture and Society* 19: 1–21.

Phillips, R. (1997) *Mapping Men and Empire: A Geography of Adventure*, London: Routledge.

—— (2002) 'Imperialism, sexuality and space: purity movements in the British Empire', in Blunt, A. and McEwan, C. (eds) *Postcolonial Geographies*, London: Continuum.

Pickles, J. (1985) *Phenomenology, Science and Geography: Spatiality and the Human Sciences*, Cambridge: Cambridge University Press.

Pile, S. (1997) *The Body and the City*, London: Routledge.

—— (1998) 'Introduction: opposition, political identities and spaces of resistance', in Pile, S. and Keith, M. (eds) *Geographies of Resistance*, London: Routledge.

Pile, S. and Thrift, N. (1995) 'Mapping the subject', in Pile, S. and Thrift, N. (eds) *Mapping the Subject: Geographies of Cultural Transformation*, London: Routledge.

Pile, S., Harrison, P. and Thrift, N. (eds) (2004) *Patterned Ground*, London: Reaktion.

Plumwood, V. (1993) *Feminism and the Mastery of Nature*, London: Routledge.

Pocock, D.C. (ed.) (1981) *Humanistic Geography and Literature*, London: Croom Helm.

Pollock, G. (1988) *Vision and Difference: Femininity, Feminism and the Histories of Art*, London: Routledge.

Pratt, M.L. (1992) *Imperial Eyes: Travel Literature and Transculturation*, London: Routledge.

Price, M and Lewis, M. (1993) 'The reinvention of cultural geography', *Annals of the Association of American Geographers* 83(1): 1–17.

Pyne, S.J. (1988) *The Ice: A Journey to Antarctica*, Iowa City: University of Iowa Press.

Rabinow, P. (ed.) (1991) *The Foucault Reader*, London: Penguin.

Raco, M. (2003) 'Governmentality, subject-building, and the discourses and practices of devolution in the UK', *Transactions of the Institute of British Geographers* NS 28: 75–93.

Rajchman, J. (1998) *Constructions*, Cambridge, MA: MIT Press.

Robinson, T. (1990) *Stones of Aran: Labyrinth*, Dublin: Lilliput Press.

—— (1995) *Stones of Aran: Pilgrimage*, Dublin: Lilliput Press.

Rose, G. (1993) *Feminism and Geography*, Cambridge: Polity Press.

—— (1999) 'Performing space', in Massey, D., Allen, J. and Sarre, P. (eds) *Human Geography Today*, Cambridge: Polity Press.

Rose, J. (1986) *Sexuality in the Field of Vision*, London: Verso.

Rose, M. (2002a) 'The seductions of resistance: power, politics, and a performative style of systems', *Environment and Planning D: Society and Space* 20(4): 383–400.

—— (2002b) 'Landscape and labyrinths', *Geoforum* 33: 455–467.

—— (2004) 'Re-embracing metaphysics', *Environment and Planning A* 36: 461–468.

—— (2006) 'Gathering "dreams of presence": a project for the cultural landscape', *Environment and Planning D: Society and Space* 24: 537–554.

Rose, M. and Wylie, J. (2006) 'Animating landscape', *Environment and Planning D: Society and Space* 24: 475–479.

Rose, N. (1996a) *Inventing Our Selves*, Cambridge: Cambridge University Press.

—— ( 1996b) 'Identity, genealogy, history', in Hall, S and DuGay, P. (eds) *Questions of Cultural Identity*, London: Sage.

—— (1999) *Powers of Freedom*, Cambridge: Cambridge University Press.

Ryan, J. (1994) 'Visualising imperial geography: Halford Mackinder and the Colonial Office Visual Instruction Committee', *Ecumene* 1(2): 157–176.

—— 1997 *Picturing Empire: Photography and the Visualisation of the British Empire*, London: Reaktion.

Ryan, S. (1996) *The Cartographic Eye: How Explorers Saw Australia*, Cambridge: Cambridge University Press.

Said, E. (1978) *Orientalism: Western Conceptions of the Orient*, London: Penguin.

—— (1993) *Culture and Imperialism*, London: Penguin.

Sauer, C.O. (1956) 'The agency of man on the Earth', in Thomas, W. (ed.) *Man's Role in Changing the Face of the Earth*, Chicago, IL: University of Chicago Press.

—— (1963a) 'The personality of Mexico', in Sauer, C., *Land and Life*, Berkeley: University of California Press.

—— (1963b) 'The morphology of landscape', in Sauer, C., *Land and Life*, Berkeley: University of California Press.

Schama, S. (1995) *Landscape and Memory*, London: HarperCollins.

Schein, R. (1997) 'The place of landscape: a conceptual framework for interpreting an American scene', *Annals of the Association of American Geographers* 87(4): 660–680.

—— (2003) 'Normative dimensions of landscape', in Wilson, C. and Groth, P. (eds) *Everyday America: Cultural Landscape Studies after J.B. Jackson*, Berkeley: University of California Press.

—— (ed.) (2006) *Landscape and Race in the United States*, New York: Routledge.

Scott, R.F. (1913) *Scott's Last Expedition*, Vol. 1, London: Smith Elder.

Seamon, D. (1979) *A Geography of the Lifeworld*, London: Croom Helm.

Sebald, W.G. (1994) *The Emigrants*, London: Harvill Press.

—— (1998) *The Rings of Saturn*, London: Harvill Press.

—— (1999) *Vertigo*, London: Harvill Press.

—— (2001) *Austerlitz*, London: Vintage.

Semple, E. (1911) *Influences of Geographic Environment*, New York: Henry Holt.

Shanks, M. and Tilley, C. (eds) (1992) *Re-constructing Archaeology: Theory and Practice*, London: Routledge.

Shanks, M. and Pearson, M. (2001) *Theatre/Archaeology*, London: Routledge.

Sheller, M. and Urry, J. (2006) 'The new mobilities paradigm', *Environment and Planning A* 38: 207–226.

Sinclair, I. (1997) *Lights Out For The Territory*, London: Penguin.

—— (2003) *London Orbital*, London: Penguin.

Smith, N. (1990) *Uneven Development: Nature, Capital and the Production of Space*, Oxford: Basil Blackwell,

Solnit, R. (2001) *Wanderlust: A History of Walking*, London: Verso.

Spinney, J. (2006) 'A place of sense: an ethnography of the kinaesthetic sensuous experiences of cyclists on Mt Ventoux', *Environment and Planning D: Society and Space* 24: 709–732.

Spirn, A.W. (2005) 'Restoring Mill Creek: landscape literacy, environmental justice and city planning and design', *Landscape Research* 30(3): 395–413.

Spivak, G.C. (1988) 'Can the subaltern speak?', in Nelson, C. and Grossberg, L. (eds) *Marxism and the Interpretation of Culture*, London: Macmillian.

Spufford, F. (1996) *I May Be Some Time: Ice and the English Imagination*, London: Faber and Faber.

Staeheli, L. and Mitchell, D. (2005) 'Turning social relations into space: property, law and the Plaza of Santa Fe, New Mexico', *Landscape Research* 30(3): 361–378.

Strohmeyer, U. (1998) 'The event of space: geographic allusions in the phenomenological traditions', *Environment and Planning D: Society and Space* 16: 105–121.

Thirsk, J. (2002) *Rural England: An Illustrated History of the Landscape*, Oxford, Oxford University Press.

Thomas, J. (1994) 'The politics of vision and the archaeologies of landscape', in Bender, B. (ed.) *Landscape: Politics and Perspectives*, London: Routledge.

—— (2004) *Archaeology and Modernity*, London: Sage.

Thomas, W. (1956) *Man's Role in Changing the Face of the Earth*, Chicago, IL: University of Chicago Press.

Thompson, G.F. (ed.) (1995) *Landscape in America*, Austin: University of Texas Press.

Thrift, N. (1996) *Spatial Formations*, London: Sage.

—— (1997) 'The still point: resistance, expressiveness embodiment and dance', in Pile, S. and M. Keith (eds) *Geographies of Resistance*, London: Routledge.

—— (1999) 'Steps to an ecology of place', in Allen, J. and Massey, D. (eds) *Human Geography Today*, Cambridge: Polity Press.

—— (2000a) 'Entanglements of power: shadows?' in Sharp, J.P., Routledge, P., Philo, C. and R. Paddison (eds) *Entanglements of Power: Geographies of Domination/Resistance*, London: Routledge.

—— (2000b) 'Still life in nearly-present time: the object of nature', *Body and Society* 6: 34–57.

—— (2001) 'Afterwords', *Environment and Planning D: Society and Space* 18: 213–255.

—— (2004) 'Intensities of feeling: the spatial politics of affect', *Geografiska Annaler Series B* 86: 57–78.

—— (2006) 'From born to made: technology, biology and space', *Transactions of the Institute of British Geographers* NS 30: 463–476.

Thrift, N. and Dewsbury, J.-D. (2000) 'Dead geographies – and how to make them live', *Environment and Planning D: Society and Space* 18: 411–432.

Till, K. (2005) *The New Berlin: Memory, Politics, Place*, Minneapolis: University of Minnesota Press.

Tilley, C. (1994) *A Phenomenology of Landscape: Places, Paths and Monuments*, Oxford: Berg.

—— (2004) *The Materiality of Stone: Explorations in Landscape Phenomenology*, Oxford: Berg.

Tolia-Kelly, D. (2004) 'Landscape, race and memory: biographical mapping of the routes of British Asian landscape values', *Landscape Research* 29: 277–292.

Tuan, Yi-Fu. (1974) *Topophilia*, Englewood Cliffs, NJ: Prentice-Hall.

Urry, J. (1990) *The Tourist Gaze: Leisure and Travel in Contemporary Societies*, London: Sage.

Valentine, G. (1989) 'The geography of women's fear', *Area* 21: 385–390.

Vasseleu, C. (1998) *Textures of Light: Vision and Touch in Irigaray, Levinas and Merleau-Ponty*, London: Routledge.

Wagner, P.L. and Mikesell, M. (eds) (1962) *Readings in Cultural Geography*, Chicago, IL: University of Chicago Press.

Wallace, E. (1993) *Walking, Literature and English Culture*, Oxford: Clarendon.

Whatmore, S. (2002) *Hybrid Geographies: Natures, Cultures, Spaces*, London: Sage.

—— (2006) 'Materialist returns: practising cultural geography in and for a more-than-human world', *Cultural Geographies* 13: 600–609.

Whatmore, S. and Thornes, L. (1998) 'Wild(er)ness: reconfiguring the geographies of wildlife', *Transactions of the Institute of British Geographers* 23: 435–454.

Wheeler, R. (1999) 'Limited visions of Africa: geographies of savagery and civility in early 18th century narratives', in Duncan, J and Gregory, D. (eds) *Writes of Passage: Reading Travel Writing*, London: Routledge.

Williams, R. (1985) *The Country and the City*, London: Chatto and Windus.

Wilson, A. (1992) *The Culture of Nature*, Oxford: Basil Blackwell.

Wilson, C. and Groth, P. (eds) (2003) *Everyday America: Cultural Landscape Studies after J.B. Jackson*, Berkeley: University of California Press.

Wilson, E. (1972) *Diary of the Terra Nova Expedition*, London: Blandford Press.

Winchester, H., Kong, L. and Dunn, K. (2003) *Landscapes: Ways of Imagining the World*, London: Pearson Prentice-Hall.

Withers, C. (1996) 'Place, memory, monument: memorializing the past in contemporary Scotland', *Ecumene* 3: 325–344.

—— (2000) 'Authorizing landscape: authority, naming and the Ordnance Survey's mapping of the Scottish Highlands in the nineteenth century', *Journal of Historical Geography* 26: 532–554.

Wylie, J. (2002a) 'An essay on ascending Glastonbury Tor', *Geoforum* 33: 441–454.

—— (2002b) 'Becoming-icy: Scott and Amundsen's polar voyages, 1910–1913', *Cultural Geographies* 9: 249–265.

—— (2002c) 'Earthly poles: the Antarctic voyages of Scott and Amundsen', in Blunt, A. and McEwan, C. (eds) *Postcolonial Geographies*, London: Continuum.

—— (2005) 'A single day's walking: narrating self and landscape on the South West Coast Path', *Transactions of the Institute of British Geographers* NS 30: 234–247.

—— (2006) 'Depths and folds: on landscape and the gazing subject', *Environment and Planning D: Society and Space* 24: 519–535.

Yeung, H. (2005) 'Rethinking relational economic geography', *Transactions of the Institute of British Geographers* 30(1): 37–51.

Zelinsky, W. (1973) *The Cultural Geography of the United States*, Englewood Cliffs, NJ: Prentice-Hall.

Zukin, S. (1991) *Landscapes of Power: From Detroit to Disney World*, Berkeley: University of California Press.

# INDEX